Environmental Requirements and Market Access

Reflections from South Asia

Environmental Requirements and Market Access

Reflections from South Asia

EDITORS:

Nagesh Kumar
Sachin Chaturvedi

Academic Foundation
NEW DELHI

First published in 2007
by :

ACADEMIC FOUNDATION
4772-73 / 23 Bharat Ram Road, (23 Ansari Road),
Darya Ganj, New Delhi - 110 002 (India).
Phones : 23245001 / 02 / 03 / 04.
Fax : +91-11-23245005.
E-mail : academic@vsnl.com
www : academicfoundation.com

Environmental Requirements and Market Access
Reflections from South Asia

edited by
Nagesh Kumar and Sachin Chaturvedi

ISBN 81-7188-466-0

Typeset by Italics India, New Delhi.
Printed and bound in India.

CONTENTS

List of Tables, Figures, Boxes and Annexures 9

Foreword by Dr. Nagesh Kumar 13

Contributors to this Volume 15

Acronyms 17

1. An Introduction
NAGESH KUMAR AND SACHIN CHATURVEDI 19

Part I
Environmental Requirements and Market Access

2. Impacts for Developing Countries
JÜRGEN WIEMANN 29

Background
Development of the Multilateral Framework of Rules and Agreements
Greening of International Trade from Below
Impacts for Developing Countries and Development Co-operation

3. Inventory of Issues
SACHIN CHATURVEDI AND GUNJAN NAGPAL 37

Introduction
Trade in ESGs
Environmental and Other Standards
Relocation of Industries
Impact of Standards and Capacity Building
Concluding Remarks

Part II
Experiences from South Asia

4. Environmental Related Barriers in the EU

S.K. MOHANTY AND T.R. MANOHARAN 61

Introduction
NTMs in the European Union and South Asia's Trade: Some Key Questions
Environment Related NTMs
Methodology
Results
Conclusions

5. Case Study of India

SACHIN CHATURVEDI AND GUNJAN NAGPAL 101

Introduction
Environment Related Trade Barriers and India
Relocation of Dirty Industries
Emerging Policy Challenges
Policy Recommendations and Concluding Remarks

6. India's Response

SHASHI SAREEN 153

Introduction
Harmonisation
Transparency
Risk-based Approach
Safety Management Systems Approach
Capacity Building
Equivalence
Rejection and Destruction of Consignments
Health Certificates
Economic Impact of Certain Measures
Conformity Assessment Issues
Conclusions

7. Case Study of Pakistan

SHAHEEN RAFI KHAN AND TOQEER AHMAD 163

Introduction
Trade-Environment Linkages in Pakistan
The Environmental Standards
The Environmental Impacts
Environmental Sensitivity of Industries
Mitigating Initiatives
Policy Options for Pakistan

8. Environmental Standards and Exports of Bangladesh

ZAID BAKHT 209

Introduction
The External Sector of Bangladesh
Impact of Environmental Standards on Exports
Shrimp Culture
Leather Tanning
Fertiliser Industry
Textile Industry and Readymade Garments
Migration of Dirty Industries to Bangladesh
Concluding Remarks

9. Case of Sri Lanka

SAMPATH JAYASINGHE 231

Introduction
Environment Management and Trade
Impact on Environment Policy on Trade
International Environmental Regulations and Competitiveness
Concluding Remarks

Part III
Policy Lessons

10. Environmental Trade Barriers and South Asian Exports: Lessons for National, International and Regional Action

NAGESH KUMAR AND SACHIN CHATURVEDI 249

The Context and Introduction
Issues for Action at the National Level
Issues for International Action
Issues for Regional Cooperation
Concluding Remarks

Index 261

List of Tables, Figures, Boxes and Annexures

TABLES

4.1 Selected Non-tariff Measures Applied by the European Union 65

4.2 Pattern of Environmental and Trade Related NTMs in the EU: Section-wise Analysis (1996) 80

4.3 Pattern of Environment and Trade Related NTMs in the EU: Chapter-wise Analysis (1996) 82

4.4 Distribution of Environment-sensitive NTMs in the EU: Section-wise Analysis (1996) 87

4.5 Distribution of Environment-sensitive NTMs in the EU: Chapter-wise Analysis (1996) 88

4.6 HS Section-wise Distribution of Average Tariff in the EU in 1996 92

4.7 HS Chapter-wise Distribution of Average Tariff in the EU in 1996 93

4.8 Pattern of Specific Tariffs and Environment-related NTMs in the EU: Section-wise Analysis 96

4.9 Pattern of Specific Tariffs and Environment-related NTMs in the EU: Chapter-wise Analysis 97

5.1 Sectoral Profile of Indian Exports 105

5.2 Top 10 Indian Agricultural Trade Partners (2001) 109

5.3 Number of Contravention Cited for US Food and Drug Administration Import Detention (August 2000-July 2001) 111

5.4 Relative Importance of Environmentally Dirty Products in World Trade 112

5.5 Growth Rates of Exports in Selected Environmentally Sensitive Products 119

5.6 FDI Inflows into the South Asian Region 123

5.7 Percentage Share of Sector-wise FDI Approvals in Total FDIs 124

5.8 Details of Various Eco-labelling Programmes 134

5.9 Market Size for Organic Products 137

7.1 Pakistan's Share in World Export 165

7.2 Environmental Hazards Associated with Textile Sector 174

7.3 Environmental Codes of Conduct 175

7.4 Environmental Hazards Associated with Leather Sector 176

7.5 Financial Implications of Pollution Charges 179

7.6 Daily Average Employment by Industry 180

7.7 Ranking Environmental Sensitivity of Industries/Products 180

8.1 Exports and Imports of Bangladesh during 1980s and 1990s 212

8.2 Changing Structure of Export (1980-81 to 1999-00) 213

8.3 Percentage Distribution of Exports by Destination 214

8.4 Price Index of Export and Import 216

8.5 Number of 4-digit H.S.-Lines under Quantitative Restrictions 217

8.6 Export of Frozen Shrimp from Bangladesh to the EU 221

9.1 Capital Cost of Effluent Standards for Wastewater Treatment 238

9.2 Estimates of Comparative Advantage by ES and ENS Products Categories 240

9.3 Overview of Direct Legislation, Relevant to Textile Products 242

9.4 Overview of Direct Legislation, Relevant to Food Products 243

FIGURES

4.1 Coverage of NTMs in the Imports of the EU 74

4.2 Composition of Frequency Ratios in the EU 75

5.1 Correlation between Pollution Intensity and Foreign Direct Investment (1991-2000) 125

BOXES

5.1 Emerging Non-tariff Barriers 106

5.2 Multiplicity of Standards 109

7.1 The Case of Kasur City 173

7.2 Chrome Recovery Plants–The Environmental Income 185

7.3 Shakarganj Sugar Mills Limited–A Zero Effluent Industry 186

ANNEXURES

A-5.1 UNCTAD Classification of Environment Related Trade Barriers 142

A-5.2 Commodities Susceptible to ETBs in India's Exports 144

A-5.3 Selected List of Disputes Referring to SPS and TBT Agreements in WTO DSU 151

A-7.1 Pakistan's Principal Textile Export 192

A-7.2 Share of Major Export Products (Rs. Million) 192

A-7.3 Western Environmental Laws on Textile Products 195

A-7.4 Western Environmental Standards on Leather Products 198

A-7.5 European Eco-labelling Scheme for Textile Products 200

A-7.6 Nordic Eco-labelling Scheme for Textile Products 203

A-7.7 European Eco-labelling Scheme for Leather Products 205

A-7.8 Brief Comparison of BS 7750, EMAS and ISO 14000 208

Foreword

The proliferation of increasingly stringent environmental requirements, food safety standards and labelling norms in developed countries has emerged as an important source of international trade frictions between the developed and the developing countries. WTO's agreement on SPS (Sanitary and Phytosanitary Measures) provides that these standards and regulations are not arbitrarily used for protectionist purposes. However, there is a widespread belief that the flexibility available in the Agreement to impose standards higher than international norms, subject to the availability of scientific evidence, has been misused by the developed countries to impose stringent regulations on the basis of very minute environmental risk assessments very often in a non-transparent and discriminatory manner. Such regulations raise the compliance costs for the exporters and may especially erode the market access of small and medium enterprises from developing countries given their limited capacity to respond to such regulations. The Doha Ministerial Conference has, therefore, instructed the Committee on Trade and Environment to give particular attention to the effect of such regulations on market access especially in relation to the developing and the least developed countries. In the light of the evidence that is now available in this and other studies, developing countries besides taking steps to strengthen their capacity to comply with such requirements, as proposed in this book, should also seek redressal of their complaints with a possible review of SPS Agreement and tightening the provisions of technical assistance in the Doha Round.

As a part of its work programme on World Trading System and Developing Countries, RIS has been looking into these concerns to assist the developing countries respond to the challenge arising from these regulations. RIS completed a study of the incidence of such regulations on exports of South Asian countries in 2002 which was undertaken in collaboration with research institutions in other South Asian countries and with the support of the UNDP under the Perez-

Guerraro Trust Fund of G-77. Subsequently, RIS organised the 'Global Forum on Environmental Requirements and Market Access: Addressing Developing Country Concerns' jointly with the OECD Trade Directorate and the Indian Ministry of Commerce and Industry in November 2002 in New Delhi. For the past couple of years, RIS has been participating in a comparative study of international food safety regulations on processed food exports in India and Thailand, conducted in collaboration with the researchers from Australia and Thailand.

The book represents in most parts the substantive output of the study undertaken by RIS with the support of UNDP assessing the incidence of these regulations on South Asian countries. The study was undertaken through a team of researchers based in different South Asian countries. The team comprised: Dr. Zaid Bakht, Director, Bangladesh Institute of Development Studies, Dhaka; Dr. Shaheen Rafi Khan and her colleagues from Sustainable Development Policy Institute, Islamabad; Dr. Mohan Manali from Nepal Forum of Environmental Journalists, Kathmandu; Dr. Sampath Jayasinghe of the Institute of Policy Studies, Colombo; besides my colleagues at RIS. The country studies prepared within the framework of the project were discussed at the workshops arranged in New Delhi by RIS. This volume contains the revised country studies and other relevant papers prepared within the framework of RIS work on these issues including those by Dr. Jurgen Wiemann, Deputy Director of the German Development Institute, and Ms Shashi Sareen, Director of the Export Inspection Council, New Delhi both presented at the OECD Global Forum on the subject organised by RIS in New Delhi in November 2002. We are grateful to all the contributors for their contributions.

I believe that the contents of the book will be found useful by policy makers, scholars and other stakeholders in agriculture, trade and industry circles, for advancing a deeper understanding of the subject.

I would like to place on record the hard work of my colleague Dr. Sachin Chaturvedi, Fellow, RIS, who coordinated this research project at RIS with his characteristic zeal.

Nagesh Kumar

Director General
Research and Information System
for Developing Countries (RIS)

Contributors to this Volume

Toqeer Ahmad .... *Sustainable Development Policy Institute (SDPI), Islamabad, Pakistan.*

Zaid Bakht .... *Director, Bangladesh Institute of Development Studies (BIDS), Dhaka, Bangladesh.*

Sachin Chaturvedi .... *Fellow, Research and Information System for Developing Countries (RIS), New Delhi, India.*

Sampath Jayasinghe .... *Institute of Policy Studies, Colombo, Sri Lanka.*

Shaheen Rafi Khan .... *Sustainable Development Policy Institute (SDPI), Islamabad, Pakistan.*

Nagesh Kumar .... *Director-General, Research and Information System for Developing Countries (RIS), New Delhi, India.*

T.R. Manoharan .... *Formerly Consultant, Research and Information System (RIS), New Delhi, India.*

S.K. Mohanty .... *Fellow, Research and Information System for Developing Countries (RIS), New Delhi, India.*

Gunjan Nagpal .... *Formerly Research Assistant, Research and Information System for Developing Countries (RIS), New Delhi, India.*

Shashi Sareen .... *Director, Export Inspection Council, Government of India, New Delhi, India.*

Jürgen Wiemann .... *Deputy Director, German Development Institute, Bonn, Germany.*

Acronyms

APTPMA	All Pakistan Textile Processing Mills Association
BIDS	Bangladesh Institute of Development Studies
CBD	Convention on Biological Diversity
CDM	Clean Development Mechanism
CITES	Convention on International Trade in Endangered Species
CREM	Consultancy on Research and Environmental Management
CP	Cleaner Production
EPR	Extended Producer Responsibility
EMS	Environmental Management System
EMAS	Environmental Management and Audit Schemes
EPA	Environment Protection Agency
EPB	Export Promotion Bureau
ESGs	Environmentally Sensitive Goods
EST	Environmentally Sound Technologies
ETBs	Environment Related Trade Barriers
ETPI	Environment Technology Programme for Industries
EU	European Union
FPCCI	Federation of Pakistan Chambers of Commerce and Industries
GMO	Genetically Modified Organisms
GPP	Greener Public Purchasing
GSP	General System of Preferences
HACCP	Hazard Analysis Critical Control Point
IISD	International Institute for Sustainable Development

MEAs	Multilateral Environmental Agreements
MRAs	Mutual Recognition Agreements
NEQS	National Environmental Quality Standards
NGO	Non-Governmental Organisation
NTM	Non-Tariff Measures
OECD	Organisation for Economic Cooperation and Development
PEPC	Pakistan Environmental Protection Council
PETF	Provincial Environmental Trust Funds
POPs	Persistent Organic Pollutants
PSMA	Pakistan Sugar Manufacturers Association
PSST	Pakistan Society of Sugar Technologists
PPPBMA	Pakistan Paper and Pulp Board Manufacturers Association
PTA	Pakistan Tanners Association
PPM	Production and Process Methods
PCP	Pentachlorophenol
PGTF	Perez-Guerrero Trust Fund
PVC	Polyvinyl Chloride
RIS	Research and Information System
SDPI	Sustainable Development Policy Institute
SMEs	Small and Medium Enterprises
SIDA	Swedish International Development Agency
SPS	Sanitary and Phytosanitary Measures
TBT	Technical Barriers to Trade
TCP	Tetrachlorophenol
UNEP	United Nations Environment Programme
UNCTAD	United Nation Conference on Trade and Development
UNCED	United Nations Conference on Environment and Development

1

An Introduction

NAGESH KUMAR
SACHIN CHATURVEDI

As the liberalisation of tariff and quantitative restrictions on trade in agricultural, food and other products has progressed, there has been an increased concern about the impact of technical measures on the exports of these products. In this regard, one important concern of the South is the stringent environmental standards imposed by the North. The technical standards such as food safety regulations, labelling requirements and quality and compositional standards have proliferated. Though empirical evidence on this in the literature is limited, some developing countries have experienced losses in exports because of difficulties to comply with certain sanitary and phyto-sanitary (SPS) measures in the import markets. These are now being termed as environment related trade barriers (ETBs).

WTO agreements on SPS and TBT aim to ensure that these standards and regulations do not cause adverse impacts on trade. Enhancing understanding of constraints faced by the developing countries in meeting the standards set by the developed countries is of key importance in the context of SPS and TBT measures by harmonising the standards. The linkages between trade and environmental measures in promoting sustainable development have been a matter of concern for the developing countries especially in the context of WTO. The extent and nature of existing regulatory structures for SPS matters in the developing countries affect their ability to comply with standards in developed countries.

The trade impacts of SPS measures can be grouped into three categories. First, they can prohibit trade by imposing trade ban on the product or on the inputs used for its production. Second, they can divert trade from one trading partner to another by laying down regulations that discriminate across potential supplies. Third, they can reduce overall trade flows by increasing costs or raising barriers for all potential suppliers. In certain cases, stricter SPS measures are applied to imports than domestic supplies. Hence, the exports from developing countries lose their competitiveness due to the higher costs they face. Environmental and health related standards and regulations in the developed country markets have the potential to create barriers to trade. The problem has become more complicated as the distinction between environmental, health and quality standards is gradually becoming very blurred. For instance, in the food sector what may be described as quality standard for food may also fall in the category of environmental standards. It is now widely believed that these technical measures impede trade of the developing countries, either implicitly or explicitly. In the Doha Declaration it is referred that issues relating to SPS measures need to be addressed on a priority basis at the next Ministerial Conference. In this regard, the Committee on Trade and Environment (CTE) was instructed to give particular attention to the effect of environmental measures on market access and trade. This has become relevant in light of the fact that the past decade has seen a global proliferation of environment and health related standards along with a rise in the trade in environmentally sensitive goods. Since the inception of WTO some 2300 notifications have been received and almost 11 per cent of them are related to environment (Nordstrom and Vaughan, 1999).

At times, it seems that, SPS measures may impede trade in agricultural and food products since in many instances they are incompatible with the prevailing systems of production and marketing in developing countries with insufficient account taken of the needs, constraints and problems faced by them. The developing countries often lack appropriate scientific and technical expertise to deal with such standards. Moreover, the multiplicity of standards in the developed country markets has further confounded the problems being faced by exporters.

Environmental and health related standards and regulations in the developed country markets have the potential to create barriers to trade. However, now it is clear that the ETBs generally cover all barriers that have been introduced by the importing country to protect the environment, as well as the health and safety of wildlife, plants, animals and humans. The developing countries have had to adjust their production processes in response to changing environmental regulations in the developed countries. Measures such as pesticide residue levels (MRL) permitted in foodstuffs, emissions standards for machines, and packaging requirements have exerted pressure on the exporters. However, what remains to be seen is the extent of impact of these measures on trade. It is now widely believed that these technical measures impede trade of the developing countries, either implicitly or explicitly.

The flexibility in WTO Agreements has been exploited by developed countries to impose stringent environmental norms and standards that are acting as significant barrier for exports of developing countries in general. These environmental standards are seen as non-tariff barriers against trade with South. For instance, adoption of a new aflatoxin standard in the European Union (EU) that would reduce health risk by approximately 1.4 deaths per billion a year while there are not a billion people in the whole of EU; African exports of cereals, dried fruits and nuts to EU are expected to decline by 64 per cent as a result of adoption of these standard with an estimated loss of export earnings of the order of US$ 670 million (Jha and Vossenaar, 2001). Another problem is the wide variation in the standards adopted by different EU importing countries. Different countries follow different norms of aflatoxins and pesticide residue increasing the compliance cost for the exporting countries.

A broader indication of impact of the SPS requirements on South Asian exports of agricultural and food products is provided by data on rejections of exports from this region. At present such a data is available only for United States. This data for August 2000 to July 2001 shows the considerable problems that South Asian countries face in meeting basic food hygiene requirements. More than 15 per cent of total agricultural imports from India and Sri Lanka were rejected because of their failure to meet these requirements (Jaya Singhe, Chapter 9). Other than that, inadequate food additives, presence of pesticide residue and heavy metals and low acid canned foods are commonly cited reasons for contravention.

There are also cases of detention of consignments of Indian spices in Spain, Italy and Germany without any satisfactory explanation forthcoming on the changes made in their regulations on microbial contamination and pesticide residue despite transparency obligation under the Article 7 of SPS Agreement. Sometimes additional standards are just imposed informally by the importing countries on their exporters as pre-conditions for doing business, as the case study of grapes exports by India has demonstrated (Chaturvedi and Nagpal, Chapter 3). Often importers use these standards to depress prices of imports. For instance, foreign buyers reportedly paid lower prices for leather products from Bangladesh because of the poor environmental conditions in the tanning sector in Dhaka (Bakht, Chapter 8). The experience of India in the case of export of flowers to Japan is also similar. Apart from this the firms also faced problems related to certification and high cost involved with it. In case of Pakistan, local firms face barriers from the certification and verification systems and complain about the high costs associated with being certified to ISO 14001 (Khan and Ahmad, Chapter 7). Many SMEs experience insufficient drivers for adoption and are uncertain about the market benefits of such systems. The lack of sector specific guidance and material tailored to different sizes of firms is an added barrier.

Some of the countries like Thailand and Brazil, which now account for over a third of agricultural exports have also faced SPS related constraints. In case of Thailand, alone, 21 SPS disputes have come up since 1995 (Mehta, 2002). Some of the cases pertaining to chicken and poultry exports to Australia and the European Union and are largely related to violation of minimum residual levels of cyanide. These levels fixed up by these countries are far more stringent than the one fixed by international agencies.

Actually in the post-Uruguay Round of trade negotiations the agriculture sector continued to be one of the most protected areas of trade especially in the European Union. The study by Mohanty and Manoharan (Chapter 4) in this volume shows that about half of total number of agricultural products and one tenth of the total manufacturing products are subject to one or another forms of NTMs. While 1910 product lines are subject to single or multiple NTMs out of 4043 agricultural products, only 1,016 product lines are subject to NTMs out of 10,208 manufacturing products.

Similarly, the LDCs also face problems in the area of sanitary regulations with the trade partners like the European Union, which had banned exports of fishery products from countries which may be affected by cholera. In 1997, such a ban was imposed on exports from Kenya, Mozambique, Uganda and the United Republic of Tanzania (UNCTAD, 1998). This problem becomes acute for countries like Mozambique, with a significant volume of exports. Consultations at the level of African, Caribbean and Pacific countries as well as on a bilateral basis have pointed to the need for measures to be implemented in Mozambique to guarantee the quality of its fishery products. The principal requirement is that inspection services must be transparent and equivalent to European Union standards. Of particular importance is the credibility of the competent authority in the exporting country. The authority must be able to demonstrate, among other things—independence of inspection services' quality control at all stages, particularly quarantine, lading and transport; and appropriate infrastructure in terms of availability of equipment, training and working conditions in the inspection services. It is, therefore, important that countries wishing to export to the European Union should invest in technical and human resources and ensure a high-quality inspection services.

More sophisticated monitoring and testing facilities, and therefore more costly procedures, are required for meeting these regulations. On top of that, the cost of rejection at the border can be considerable, as it includes loss of product value, transport and other export costs and product re-export or destruction. A number of agricultural products of South Asia are facing SPS related problems. In case of products like peanuts, other nuts and milk, EC has introduced high level of protection by reducing the maximum level of presence of aflatoxin in these products. The level of protection proposed by EU is substantially higher than that provided under Codex recommendations. In the case of peanuts, the EU argument has been that the risk involved is of persons contracting cancer in a population of one billion. This is extremely unreasonable because EU population is less than one third of a billion. So the level of SPS protection is not in relation to the extent of risk involved.

This volume examines the incidence of these emerging trade barriers on exports of South Asian Countries.

Layout of the Volume

The subject matter of this volume is divided into three parts. The first part covers an overview of broader issues related to the environmental requirements and market access and reviews of literature. Wiemann (Chapter 2) discusses the broad setting in context of developing countries. Chapter 3 by Chaturvedi and Nagpal provides an inventory of issues in the current debate on environmental standards and trade with specific focus on environmentally sensitive goods.

The part two of the book attempts to put together national experiences from South Asia. The country studies cover India, Pakistan, Bangladesh and Sri Lanka. Chapter 4 by Mohanty and Manoharan on environment related non-tariff measures faced by the South Asian countries in their trade with EU brings out that the EU market is largely protected by the NTMs. The authors have used frequency-ratio to examine the level of NTMs protection. They find that in most agricultural products NTMs are high. The case study on trade and environmental standards and migration of dirty industries in India by Chaturvedi and Nagpal (Chapter 5) puts together evidences on how environmental standards are affecting agricultural exports and also brings out the need for domestic preparedness to face these challenges.

Sareen (Chapter 6) presents the Indian response to the changing standards and quality issues in international trade. Khan and Ahmad (Chapter 7) discuss the trade and environmental standards in context of Pakistan. They suggest that there is a need for mutual appreciation of constraints by North and South so that the environmental standards debate does not remain confined to polarities. This Chapter also presents data on the migration of dirty industries to Pakistan.

Bangladesh has been working out various strategies to cope up with the environmental standards without allowing them to adversely affect the exports of Bangladesh. As discussed in Chapter 8 by Bakht, it becomes clear that Bangladesh needs a planned and consistent technical assistance to cope up with environmental standards. In this regard, the Chapter has reviewed select export items from Bangladesh such as shrimp, leather, and fertilisers. Jayasinghe (Chapter 9) discusses the trade and environmental standards debate in Sri Lanka and suggests a strategy to tap the first mover advantage of Sri Lankan exporters by an early implementation of technical assistance programmes for improving the standards.

The broad policy recommendations of this book are encapsulated in the last Chapter by Kumar and Chaturvedi. It takes account of the implications at the national, international and regional level for various agencies including national governments and international agencies to help South Asia address the environmental trade barriers and their implications on regional exports.

References

Jha, Veena and René Vossenaar (2001). *Strengthening Developing Countries' Capacities to Respond to Health, Sanitary and Environment Requirements: A Scoping Paper for South Asia*, UNCTAD, New Delhi.

Mehta, Rajesh (2002). "International Food Safety Regulations and Processed Food Exports from India: An Exploration into Research Agenda", *Launching Workshop Proceedings*, June, RIS, New Delhi.

Nordstrom, Hakan and Scott, Vaughan (1999). "Trade and Environment", *WTO Special Studies* 4, WTO, Geneva.

UNCTAD (1998). *The Least Developed Countries: Report*, UNCTAD, Geneva.

Part I

Environmental Requirements and Market Access

2

Impacts for Developing Countries

JÜRGEN WIEMANN

Background

About some years ago, I did a study with a team of young trainees from the German Development Institute (GDI) on the impact of new environmental standards and requirements on three Indian industries, textiles/clothing, leather and refrigerators. Two of these industries were experiencing for the first time the impact of new environmental product standards on their European export markets, whereas the third had to adjust to a multilateral environmental agreement (MEA), the Montreal Protocol, stipulating and offering technical assistance to the phasing out of CFC, the chemicals causing damage to a global public good, the ozone layer. Our empirical study had two aims: (1) to raise awareness among the Indian business community and the Indian commerce and industrial policy authorities of the new requirements for exporting and to ask how they would be able to comply with environmental regulations and standards and, (2) to report back to the German aid community that there would be an increasing need for technical assistance to exporting industries and to the whole institutional infrastructure for testing of export goods and certifying their compliance with the environmental market requirements.

Development of the Multilateral Framework of Rules and Agreements

Trade and environment has become a booming 'industry' in development co-operation since then, not least thanks to the activities of both the OECD and the UNCTAD Secretariat. The multilateral rules-based system for international trade has been responsive to the new theme of environmental and health standards and produced some new rules for dealing with trade conflicts in this field. In reaction to the 1992 UN Conference on Environment and Development (UNCED) in Rio de Janeiro, the concept of sustainable development enshrined in the Rio Declaration and Agenda 21 was woven into the Preamble of the new World Trade Organization (WTO) into which the old GATT Secretariat had been converted at the end of the Uruguay Round in 1994. A Committee on Trade and Environment (CTE) was established within the WTO to sort out the problems and to suggest solutions for trade conflicts with an environmental background. The GATT Agreement on Technical Barriers to Trade (TBT) was refined and adjusted to the new requirements and, even more important, a new agreement was added, the WTO Agreement on Sanitary and Phytosanitary Measures (SPS) that contained special provisions for trade-relevant regulations and other regulations on animal and plant health that could become trade barriers.

Based on these new multilateral rules and agreements, a number of trade conflicts between WTO members have been brought to the new and strengthened WTO dispute settlement procedure, and some decisions of Panels and the Appellate Body have established and broadened the case-law basis for future rule-making and interpretation in the field of trade and environment. Most famous are the GATT and later WTO decisions on the tuna/dolphin and the shrimp/turtle disputes between the United States and some developing countries. These decisions have underlined that trade measures can be used for environmental purposes, but only under certain clearly defined circumstances and requirements. They can only be used if a global public good is involved like the endangered species sea turtles which is protected by multilateral agreements like the Biodiversity Convention or CITES. And the trade measures must not discriminate between exporting countries, and technical assistance should be provided to developing countries for assisting producers, exporters and government agencies to comply with the goal of protecting endangered species or achieving other global environmental goals.

Other decisions of the WTO dispute settlement procedure are related to SPS conflicts. Interestingly, the three cases settled so far are conflicts between developed countries. This shows that trade conflicts over different perceptions of environmental and health risks are not a North-South issue but arise between developed countries, and even within each country there are conflicts between advocates of free trade and environmental activists. Most famous is the dispute between the US and the EU on the European import ban of hormone-treated beef. This conflict is based on the new and extremely important provision of the SPS Agreement that trade measures taken to prevent health risks for consumers have to be justified with scientific evidence of the risks and scientifically founded risk analysis. The EU has not yet produced this evidence but nevertheless maintains its import ban by claiming that the precautionary principle gives the authorities of the importing country the right to protect its citizens against health risks arising from imported goods. It would be too late, the EU argues, if scientific tests would prove, only years later that the substance in dispute does cause health damages. Of course, domestic farm interests of the EU are involved in this case, but it is difficult to separate them when they are backed with health arguments shared by a majority of the citizens.

After several food scandals, for e.g., the mad cow disease or the foot and mouth disease which led to the reintroduction of intra-EU trade restrictions to avoid importing health risks, consumers are alert and ask for more guarantees for the safety of their food. An increasing number of Europeans feel that the EU standards for food are still too low and that compliance is not enforced effectively. In Germany, a new NGO, Food Watch, is coming onstage which will put pressure on the legislature to introduce stricter standards and will monitor how the authorities are controlling and enforcing them properly. Developing countries exporting food to the EU or other OECD countries must be aware of the fact that the consumers demand from their governments and food inspection bodies to enforce compliance regardless of the countries of origin. Consumers will not make concessions in the area of real or felt risks for their health and, to a lesser degree, for the environment even if they are told that developing countries' exports will be affected. The next conflicts are already visible, for e.g. the conflict over different perceptions of health risks of genetically modified (GM) food between, for e.g., the EU and the USA.

One example for a trade measure for environmental protection and risk avoidance completely dissociated from protectionist motives is the conflict between Thailand and Australia over durian exports. Australia argues that its precarious ecological system could face an incalculable risk from pest or other diseases imported with durian from Thailand or certain areas of that country. There is no durian production in Australia to be protected, and the potential consumers of durian in Australia will mainly be immigrants from Southeast Asia who should be able to calculate their health risk themselves. Nevertheless, the risk of an epidemic spread of diseases or other new organic material unknown to an especially vulnerable ecosystem like Australia, delinked from the rest of the world can be calculated higher than the damage associated with an import restriction.

It seems as if there is an impasse between those who claim that trade measures may only be taken once scientific evidence has been produced and those who refer to the precautionary principle and claim the right to prevent imports as long as their consumers feel a risk even without scientific evidence. One way out of this impasse is product labelling that gives the consumers—instead of governments—the right to discriminate between different suppliers of their food stuff. It is difficult to argue against strengthening consumer sovereignty in an era in which more and more people around the world feel threatened in one way or the other from globalisation, however justified these concerns may be. Product labelling had been endorsed by the GATT dispute settlement decision on the tuna/dolphin case. It said that the US was not allowed to impose a general import ban, but that importing companies were allowed to put a label on the tuna products stating that their tuna had been caught with dolphin safe nets.

Another option might be, is to get insurance companies involved in what they specialise, i.e. risk assessment and insurance against risks. Australia could ask Thailand to insure its durian shipments against the risk of causing damage to Australia's ecological system, the EU could ask the US to find an insurance for its beef exports to Europe that would cover the health risks that Europeans believe to be involved with eating hormone-treated beef or, in future, genetically engineered food. If the risks involved are assessed to be within the usual range, insurance companies should be willing to cover these risks. If they find the risks to be too high and unpredictable, their reluctance to cover these risks would be an indication that in this case the EU is right to base its trade measures on the precautionary principle.

Greening of International Trade from Below

Even more important than the changes on the government level and within the multilateral trading system are the changes that take place on the market level. Here, environmental standards introduced by companies, associations and NGOs are gaining importance. They can be described as an instrument of product differentiation. Trading differentiated products is different from trading homogeneous products, in which comparative advantages based on cheap labour or lower national environmental standards play a decisive role. Therefore, textbook economics advocates lower environmental standards for developing countries as another comparative advantage of these countries in international trade. From this reasoning there is little justification for raising environmental standards in developing countries, and persuading them to do so might be seen from the free trade viewpoint as just another protectionist strategy in disguise.

In real-world trade, however, differentiated products play a greater role compared to homogenous goods. Design, quality and other aspects that determine consumers' buying decisions play a more important role than small price differentials on which standard economic theory is focussed. Brand names for the products of transnational companies guaranteeing consistent quality and services to consumers can help to increase world market shares of a product and a company. In order to guarantee consistent quality, the whole production/supply/marketing chain has to be organised by the company or an association of companies that is behind the brand name. And with the progressive dismantling of trade barriers as a result of the past GATT and WTO rounds, integrated production and marketing chains can be organised across borders ever more easily. Developing countries' producers linking to such production and marketing chains feel the impact through increasing quality requirements they have to fulfil if they want to stay in the chain as reliable suppliers.

Industries like textiles and clothing which are not dominated by multinational corporations with their own brand names, can imitate this pattern with some modifications. Smaller companies can form associations agreeing on certain quality standards—for example, environmental-quality standards and guarantee consistent (environmental) quality with an (eco) label (e.g. ECOTEX 100 in Germany). These associations introducing certain product or even PPM standards to be guaranteed by their label are gaining importance in international trade. The more affluent consumers in the developed countries—but increasingly also in developing countries—are prepared to pay a higher price for goods that meet these environmental requirements. Even if they do not pay a higher price, but prefer labelled products to others, they put the producers and exporters of the former at an advantage through increasing their market shares. This trend can be clearly seen for food products and for textiles and clothing.

If standards like measures for length, weight, etc. can be described as public goods, standards established by eco-labels for a loose association of producers—and consumers paying the higher prices as entry fee to the club defined by an eco-label—can be described as club goods. A special feature of these clubs can be that the standards they establish do not only contribute to the utility of the club members, but can also contribute to the production/protection of global public goods, such as the prevention of the destruction of the ozone layer or the prevention of climate change in general.

If that is the case, these associations/clubs experimenting with new eco-labels and ways and means to enforce them along international production/supply/marketing chains should be encouraged and supported by governments who are already relying on civil society activities for solving public problems in other areas. Governments can help, e.g., to counteract the proliferation of eco-labels that jeopardises the credibility with consumers by encouraging certain basic standards on which all labels can agree so that sooner or later there will be only one or a few labels for one product group. Governments can also help to establish and improve the institutional infrastructure for testing the products, monitoring production processes (e.g. organic food) and certify the quality for the agency issuing the label. And governments can act as 'certifiers of the last resort' monitoring the credibility of labels and private certifying agencies and their capacity to do accurate testing, etc. By that, governments can prevent unfair competition of companies using labels that are not based on sound criteria, effective testing and credible certification.

Impacts for Developing Countries and Development Co-operation

Developing countries' governments have to:

- search for and disseminate information among export-oriented producers about environmental requirements on export markets;
- help their producers to adjust products and PPMs for compliance with standards on export markets legislate and enforce appropriate and effective domestic environmental and health regulations;
- play an active role in international standardisation bodies,
- establish testing and monitoring institutions, and have them accredited by international bodies or respective institutions in the target countries;
- keep an eye on whether importing countries are using their environmental and health regulations in a discriminatory way, and raise their voice in bilateral negotiations or in the WTO if these regulations create unnecessary obstacles to trade; and
- last but not the least ask for technical assistance from bilateral and multilateral aid agencies.

Development assistance by multilateral and bilateral aid agencies can and should support all these activities of developing countries, especially by providing early warning information about upcoming new standards and regulations, and help to establish the necessary infrastructure for testing and certification. Developed countries have promised many times in WTO to assist developing countries to adjust to the new WTO agreements, e.g. the TRIPs Agreement or the TBT and the SPS Agreements. But, they have not increased their aid budgets to fulfil these new requirements for trade-related development assistance so that there can be a trade-off between the new programmes needed for trade development and traditional poverty alleviation programmes. Aid agencies should be consistent when designing and implementing new trade promotion programmes. They should not promote new export crops before having assessed the present and future environment requirements on their own markets which might jeopardise market access of the new crops promoted (*cf.*

the snow peas case in Guatemala documented in one of the papers submitted to this conference).

Finally, bilateral aid agencies should take up the task of informing consumers and the general public in their own countries about the role they play—and could play with larger aid volumes—in raising environmental awareness and standards in developing countries and by this help to reduce the tensions between developed and developing countries over mutual allegations of eco dumping *versus* environmental protectionism. Thus, they could contribute to take some pressure off the WTO to get too deeply involved in environmental matters for which WTO has neither a mandate nor a comparative advantage.

3

Inventory of Issues

SACHIN CHATURVEDI
GUNJAN NAGPAL

Introduction

This chapter is an effort to inventorise the various issues in the trade and environment debate in context of trade in environmentally sensitive goods. The focus of the chapter is to study the impact of increased product based standards on the trade of South Asian countries and to examine if there is any migration of environmentally sensitive industries into South Asia.

Trade in ESGs

The term Environmetally Sensitive Goods (ESGs) has been defined differently by different authors. According to the conventional approach, industries that have incurred high level of abatement expenditure per unit output are identified as dirty industries (Tobey, 1991; Low and Yeats, 1992). Another approach is to select sectors, which rank high on actual emissions intensity (emissions per unit output) (Mani and Wheeler, 1998). The third and least commonly used approach is based on the effects of the product on large-scale conversions or degradation of natural areas or loss of biodiversity. In this case, the product may not necessarily create any environmental damage during its consumption or disposal in the importing country but it can cause long-term economic losses in the exporting country through irreversible damage to the ecosystem.

Recently widespread concerns have been expressed about the relationship between international trade of environmentally sensitive goods and environmental regulations. Various authors have made effort to examine whether the pattern of export performance of environmentally sensitive goods underwent systematic change. Low and Yeats (1992) examined the trade in ESGs for selected developed and developing countries for the years 1965 to 1988. The study covered a few ESGs. According to them, industrial countries accounted for approximately three quarters of trade in these products, with almost 40 per cent of world exports originating from EC. In contrast, developing countries account for only about quarters of global exports of ESGs, with about one third of these shipments originating from Southeast Asia. Low and Yeats (1992) also examined the secular global trend of trade in ESGs and found that while there was a drop of six percentage points in the value of shipments originating in North America, the share of Southeast Asia more than doubled.

A similar study was done by Xu Xinpeng (1999) for 34 countries, which included 25 OECD countries and some developing countries. The idea of the study was to get a full picture of the changing trade patterns of ESGs for year 1965-98. The 34 countries accounted for nearly 80 per cent of world exports in ESGs in 1995. It was, therefore, to be expected that this analysis would provide a full picture of changing performance of ESGs over time. Two different means to break down the export performance data of ESGs, using a histogram and time series pattern, were employed in the study. The study concludes that export performance of ESGs for most countries remained unchanged between the 1960s and the 1990s, despite the introduction of stringent environmental regulation in most developed countries in 1970s and 1980s. However, a similar study done for the South Asian countries by Chaturvedi and Nagpal (2001) shows that while the environment related standards are proliferating, at the same time the trade in ESGs is also increasing. Their study empirically shows that there is a significant rise in such exports originating from the South Asian countries.

This is in contradiction to Sorsa (1994) who comments that the world market shares in environmentally sensitive goods have not changed dramatically over past two decades, despite the introduction of higher environmental standards in most industrial countries. He pointed out that over 70 per cent of world imports of environmentally sensitive goods continue to originate in industrial countries. Similarly, Tobey (1991) sets up a Hecksher-Ohlin Vanek (HOV) multifactor, multicommodity model. Tobey regresses the net exports of five pollution intensive industries on the stocks of productive factors, including the environment. He found no evidence that the introduction of environmental control measures has caused trade patterns to deviate from the HOV predictions. Grossman and Krueger (1993) investigate empirically the environmental impacts of North America Free Trade Agreement (NAFTA). They find that though traditional determinants of trade are significant, the alleged competitive advantages created by lax pollution controls in Mexico play no substantial role in motivating trade flows.

Another important aspect of trade in ESGs debate is the argument related to the Multilateral Environmental Agreements (MEAs). The debate on the role of MEAs *viz-à-viz* trade has been highly polarised.[1] However, authors like Rutgeerts (1999) have found it reconcilable by stating that free trade may lead to such negative effects (externalities) as described by environmentalists only if appropriate environmental polices are not pursued. MEAs are considered to have trade and competitiveness effects. However, Sorsa (1992) observes that it is difficult to impose any kind of trade restrictions against any member or a non-member violating these treaties in the present format of international trade rules.

There are very few studies which have made an effort to quantify the implications of international treaties such as the Basel Convention; The Convention on International Trade in Endangered Species (CITES) or the Montreal Protocol on trade of developing countries. According to Jha *et al.* (1999) the effects are more significant on the developing countries because the overall difference in income levels would make the relative cost of adjustment more expensive for the developing nations. Also, the growth and the export pattern of countries tends to be such that during the early stages of industrialisation they may be more reliant on the export of products, which are banned or restricted by MEAs. CITES regulates commercial trade in endangered and other species. CITES could also result in unintended trade and competitiveness effects on some parties. Zimbabwe, for example, as a party to CITES, reports significant trade losses with regards to crocodile farming and the ivory trade. Zimbabwe is said to have an

1. Anderson *et al.* (1995).

estimated stock pile of US$12 million worth of ivory that it cannot legally sell. Intal (1994) has shown that timber exports by the Philippines could be significantly affected by the agreements based on principles of sustainable development.

Environmental and Other Standards

The issue of environmental regulation and international competitiveness has encouraged the debate on harmonisation of standards. According to Pearson (2000) the major objection to internationally harmonised or uniform standards is that economic, social and environmental conditions differ from one country to another, and what is appropriate in one set of circumstances may be inappropriate in another. Bhagwati (1996) discusses that if harmonisation on policies among nations is a pre-condition for free trade. Leebron (1996) provides justification for harmonisation of standards for free trade. The study points out that harmonisation between two jurisdictions might appear necessary to make certain transactional activities possible, but in most cases it will simply be more efficient. However, according to Esty and Geradin (1998), the belief that harmonisation and imposition of welfare-reducing uniform environmental standards are the only regulatory tools available to address competitive issues is wrong.

The emergence of global marketplace has resulted in debate on the possible impact of environmental standards and regulations on national trade competitiveness. In the developed countries, the major worry is that the introduction of more stringent environmental regulations in the domestic market would result in loss of competitiveness *vis-à-vis* industries in countries where standards are less stringent. Developing countries, on the other hand, are concerned that increasingly stringent regulations in developed countries could adversely affect market access for their export products.

Markandya (1999) discusses the link between trade and the environment related standards and suggests that developing countries have to adjust their production process in response to changing environment requirements in the developed countries. The case in point is that of meat and drug production facilities, which European importers emphasise to visit. This gives chance to get involved in the process of production and is against the spirit of Article III of GATT which forbids imposition of process standards on trading partners. These issues are discussed in the literature giving special attention to developing and transition countries where the relationship between trade and the environment has a major impact.

It is indicated that effect of environmental requirements on trade and export competitiveness of developing countries depends on factors, such as destination of exports, cost structures, basis for export competitiveness, relationship with foreign firms, availability of raw materials and other inputs, technology, and information. The study points out that firm size is of particular importance. The environmental effects of changes in production patterns in developing countries resulting from trade liberalisation differ on all the three counts, *viz*. country, sector and commodity. Pearson (2000) approaches the competitiveness question from the perspective of advanced industrial countries, and mostly use US pollution or pollution-abatement cost data. The principle conclusion is that there is little or no evidence that strict environmental controls result in any significant loss of international competitiveness.

The new agreements on TBT and SPS measures were added to the WTO, with an idea that no country should be prevented from taking measures necessary to ensure the quality of exports, or for the protection of human, animal or plant life, or health of the environment, or for the prevention of deceptive practices, at the level it considers appropriate. However, barriers in the name of technical regulations line the boundaries of internationally trading nations. The TBT and the SPS measures have been a major source of non-tariff barriers. Various studies have addressed the issue of SPS measures and developing country exports. More theoretical work has demonstrated that developing countries find it difficult to trade with the developed countries due to differences in quality requirements, which in turn reflect prevailing consumer or the nature of government regulation (Murphy and Shleifer, 1997). Saqib (2000) examines some examples of measures perceived as non-tariff barriers by the Indian exporters, which are disguised under the logic based on various grounds like quality, manufacturing process, certification, testing methods, environment, etc. However, some of the studies like CUTS (1995) have been focussed on sectoral implications of standards. This study particularly looked into textiles and garments.

Maskus *et al.* (2000) have evolved a framework to analyse the quantification of trade barriers in terms of their impact. The paper also outlines the strengths and weaknesses in Mutual Recognition Agreements (MRAs). This is a model of regulatory harmonisation first developed as part of internal market reform in the European Community in late 1980s. Some application on this line has come from Otsuki *et al.* (2000), who have used gravity equation model to estimate the impact of changes in food related EU standards on African exports. They did a survey of trade and regulatory data for 15 European countries and 9 African countries between 1989-1998. The results suggest that the implementation of the new aflatoxin standards in the EU will have a negative impact on African exports. The EU standards, which would reduce health risk by approximately 1.4 deaths per billion a year, will decrease the African exports by 64 per cent or US$ 670 million. Apart from this very few studies have made an effort to quantify the impact of SPS measures on trade of developing countries. Cato (1998) attempts to quantify the costs of compliance with SPS measures by developing countries. This study assesses the costs of upgrading sanitary conditions in the Bangladesh frozen shrimp industry to satisfy EU and US hygiene requirements. It is estimated that $ 17.6 million was spent to upgrade plants over 1997-98. This gives an average expenditure per plant of $283,000. The total industry cost required to maintain HACCP is estimated to be $ 2.2 million per annum. Finger and Schuler (1999) examined the costs of SPS requirements in the developing countries. They found that the cost of achieving disease and pest free status to enable Argentina to export meat, fruit, and vegetables is reported to have been $82.7 million over the period 1991-1996. Mutasa and Nyamandi (1998) assess the degree to which SPS requirements impede exports of agricultural and food products from African countries. Of the African countries, 57 per cent indicated that exported products had been rejected within the previous 2 years. The main reason was microbiological contamination.

Some standards evolved in the developed countries, as voluntary benchmarks for importing goods have become mandatory standards for the developing countries. Eco-labelling is one of them. It has generated many concerns in developing countries. They argue that the voluntary eco-labelling programmes are of discriminatory nature and are, thus, inconsistent with free trade principles of the WTO (Tietje,

1995). Since they are designed to differentiate products on the basis of their environmental features, they can have a major influence on conditions of competition in a market. Adverse trade effects may arise, first of all, from lack of transparency of product selection, criteria development and threshold-setting processes. Exporters may have difficulty in gaining information about these measures. In addition, the process of obtaining labels or ensuring compliance may pose greater difficulty for foreign producers. Moreover, there is the potential for market fragmentation arising from various eco-labelling schemes (Chang, 1997). Eco-labelling may also have an impact on cost competitiveness and on the attractiveness of the product in the market. (Jha and Vossenaar, 1999).

A case study on Malaysia by Markandya (2000) states that mandatory and voluntary labelling requirements have had an adverse effect on exports though the study does not give an estimate of the size of this impact. Another study on Thailand estimates the impact of eco-labelling schemes as negligible in the key markets such as North America, East Asia and Europe. The study notes, however, that in the case of one eco-label developed in Germany to apply to textiles, Thai producers are adversely affected. Wyatt (1997) discusses the consequences of a number of voluntary environmental measures on the exporters of developing nations.

However, certain developing countries have attempted to absorb these voluntary standards as official standards for exports; for instance, in India, the government has launched an Eco-labelling scheme for various products. However, a study by Jha *et al.* (1999) reports that Indian exporters are not interested in subscribing to international eco-labelling schemes. These studies show that most producers do not feel that the evolving eco-labelling schemes have affected their exports significantly. However, it is widely acknowledged that these measures could potentially have a detrimental effect on the exports of developing countries.

Various cases have been brought under the WTO's new Dispute Settlement Understanding with a focus on SPS and TBT dimension. Hufbauer *et al.* (2000) discusses some SPS disputes that have gone through the entire WTO Dispute Settlement Mechanism in 1999. The cases discussed include EC–beef hormones from the US and Canada; Australia–salmon from Canada; and Japan–varietals from the U.S. In the Australia- salmon from Canada, Australia continues its 1960s era ban against the import of fresh salmon from Canada, citing old evidence. The Appellate Body, after a detailed analysis, found that the ban was not consistent with the level of sanitary and phyto-sanitary precautions applied to domestic salmon. The WTO website provides the details of various trade related environmental disputes. One of the TBT cases is the 'shrimp-turtle' case brought by India, Malaysia, Pakistan and Thailand against the US. In early 1997, a joint complaint was made against a ban imposed by US on the importation of certain kind of shrimp and shrimp products. Under the US Endangered Species Act of 1973, the US required its shrimp trawlers use 'turtle excluder devices' (TEDs) and hence, the exporters had to follow similar requirements. It banned exports from the South Asian countries and provided countries in the Western Caribbean technical and financial support and longer transition periods. The US lost this case because it was discriminating between WTO members on pretext of environment protection.

Robertson and Kellow (2000) explore the aspects of risk with special reference to the WTO, where national instruments to reduce risks may conflict with international trade rules. Quarantine regulations, technical/product standards and environmental legislation in some circumstances may conflict with trade rules and principles, and result in trade disputes. WTO treatment of risk is also important in the context of agreement on SPS and TBT. The authors conclude that the WTO has no role in assessing scientific risk, especially when environment is involved.

Similarly, there are a number of issues that arise in the context of WTO rules with respect to genetically modified products and their labelling. The first is whether naturally grown fruits and genetically engineered fruits should be regarded as 'like products' in the context of WTO rules. One argument advanced is that as genetically engineered fruits may not reproduce at the end of their life span, the final product characteristics are affected by the process and production method. Therefore, they cannot be regarded as 'like products'. At the same time as the end use of GMOs may be same as naturally grown products, they could be regarded as 'like products'. Another related issue that is particularly relevant in the context of the Biosafety Protocol relates to the labelling of products using GMOs. While some countries believe that such labelling would be entirely consistent with WTO rules on the grounds of public health concerns, other countries argue that their exports may be affected by such labelling.

Relocation of Industries

Another concern is that differences in the stringency of environmental process standards may induce industries located in high-standard jurisdiction to relocate to low standard jurisdiction. A number of studies are available in the literature on the impact of environment on industrial relocation. Majority of the studies addressed the issue in the context of developed countries, particularly United States. These studies have generally defined the polluting industries on the basis of pollution abatement cost and other expenditures. The definition of dirty industries is crucial to the analysis, but varies from one study to another. This problem is not that serious, as several different definitions seem to yield similar lists of dirty industries. Table 3.1 shows the industries that are identified as dirty industries by various authors.

Kalt (1988) was one of the first to test the 'pollution haven' hypothesis. He examined US net exports in 1977 and studied the relationship between the level of compliance costs and the change in net exports. He concluded that there is some evidence of a negative effect of environmental regulation on the US net exports (a validation of the 'pollution-haven' hypothesis), but the relationship is statistically weak. Han and Braden (1996) expanded and updated the Kalt study, examining 19 manufacturing industries in the US between 1973 and 1990. They looked for the relationship between pollution abatement costs and net exports over time. Their regression results show pollution abatement expenditure having statistically significant negative effect on net exports. Han and Branden also analysed the elasticities of net exports with respect to pollution abatement expenditures for the 19 industries. Large elasticities would imply that added abatement expenditure would lead to a substantial marginal decrease in net exports. The industries shown to have large elasticities and large abatement costs were paper and allied products, chemicals and primary metals. Many industries, including furniture, printing, leather and allied products, fabricated metals, petroleum and coal products, had elasticities close

Table 3.1

Identification of Dirty Industries across Studies

Author	*Dirty Industries as Identified by the Author*
Kevin Gallagher and Frank Ackerman	Iron & Steel; Non-ferrous Metals; Industrial Chemicals; Pulp & Paper; Non-metallic Minerals. (3-digit SITC level)
Raman Letchumanan	Chemicals; Primary Metals; Paper; Plastics; Fabricated Metals; Leather.
Low and Yeats	Iron & Steel; Non-ferrous Metals; Refined Petroleum; Metal Manufactures; Paper and Articles.
The World Bank (1998b, p.113)	Iron & Steel; Non-ferrous Metals; Industrial Chemicals; Petroleum Refineries; Non-metallic Mineral Products; Pulp & Paper Products.
Tobey	Mining; Primary Metals; Paper & Pulp and Chemicals. (3-digit SITC level)
Xinpeng Xu	Iron & Steel; Cement; Chemicals; Newsprint Paper; Non-ferrous Metals.

Source: RIS Discussion Paper # 22.

to zero. Over the 18-year period, the elasticities declined in almost all industries. The author concludes overall that there has been a negative effect on net manufacturing exports due to environmental regulations. Letchumanan (1998) empirically tested the 'pollution-haven' hypothesis by analysing the current pattern of cross-border industrial location. This was done by correlating FDI flows with pollution intensity of each industrial sector, with the pollution intensity of each industrial sector, for a number of developed and developing countries. The results proved beyond doubt that the 'pollution-haven' hypothesis is not empirically valid.

A 1991 OECD analysis observed that some shift of competitive advantage to countries with lower environmental standards in pollution-intensive textile and leather tanning industries had occurred. A 1993 OECD study concluded that some environmentally dirty activities, particularly in the resource-based sectors, had migrated to lower income countries with weaker environmental standards, and that the result was a geographical shift in production capacity within sectors with a consequent acceleration of industrial pollution intensity in developing countries. Xing and Kolstad (1996) conducted an empirical study to examine the effect of the laxity of environmental regulation on FDI. They examined foreign direct investment in 22 countries by the US chemical industry, the US electrical industry and the US non-electrical industry in 1985-1990. The authors use measured sulphur emissions as the proxy for stringency of environmental policy, reasoning that there is a direct relationship between policy measures and the measure of pollution. Their statistical analysis shows that the laxity of environmental regulations in a host country is a significant determinant of FDI from the US chemical industry, though not from less polluting industries. Furthermore, the more lax the regulations in the host country, the more likely it is to attract the investment capital of the US industry. Eskeland and Harrison (1998) investigate whether inward FDI in developing countries is concentrated in polluting industries. The study covers investments into Mexico, Venezuela, Cote d'lvoire and Morocco during the 1980s, with the first two countries receiving most of their investments from United States and other two from France. No evidence was found to suggest that the investments in these countries are biased towards polluting industries.

Low and Yeats (1992) analysed the effect of environmental policies on the industrial location. The analysis concludes that the increase in developing countries Revealed Comparative Advantage (RCA) was far greater than that of the industrial countries and dirty industries account for an increasing share of export of developing countries.

Lucas *et al.* (1992) analysed the data for 80 countries from 1960 to 1988 to examine how the relocation of dirtier industries do occur at least on the basis of broad differences in trends of environmental regulation across differing time periods. Results suggest that stricter regulation of pollution-intensive production in OECD countries have led to significant locational displacement, with consequent acceleration of industrial pollution intensity in developing countries. However, the authors pointed out that the observed phenomenon merely reflects dispersion or industry expansion, as opposed to displacement. The paper finds only a composition effect, so there is no evidence that the industry has been uprooted from industrial countries.

A World Bank research (Mani and Wheeler, 1997) extended this research further using international data on industrial production, trade, and environmental regulation for 1960-1995. This research examines shifts in trade and production patterns in the main trading regions of the world economy in relation to changes in other factors, such as income growth, land prices, energy prices, and environmental regulation over the 35-year period. The authors conclude that their cross-country analysis has found a pattern of evidence, which does not seem consistent with 'pollution haven' hypothesis. Pollution-intensive output as a percentage of total manufacturing has fallen consistently in the OECD and risen steadily in developing world. Moreover, the periods of rapid increase in net exports of pollution-intensive products from developing countries coincided with periods of rapid increase in the cost of pollution abatement in the OECD economies.

Abhimanyu (1996) examined the different ways in which international trade and trade agreements affect the environment of a country within a regional co-operation effort, and the ways these effects pertain to international environmental policy. The focus is on regional APEC. Results show that dirty products have expanded faster in developing countries than the average of all industrial countries over the past decade.

Pearson (1987) surveyed several studies, all of which tend to support the conclusion regarding insufficient evidence for industrial flight to developing countries. Leonard and Duerksen (1980) determined whether ECC differentials have led to industrial flight towards LDCs by employing trade and investment data. The important results that came to light were: host countries which received the most overseas investments in pollution-intensive chemical, paper, metals and petroleum refining were other industrial countries (not LDCs) and the percentage of the US, FDI in pollution-intensive industries in LDCs compared to DCs did not increase significantly. Leonard (1988) presented case studies of FDI in Ireland, Spain, Mexico and Romania arguing that the industrial flight and pollution-haven hypothesis, are based on static idea of comparative advantage. Examining aggregate trade and investment statistics, Leonard sees no evidence of large-scale industrial flight as a response to US environmental regulations.

The available studies generally agree with the fact that the dirty (polluting) industries have grown faster in developing countries. South Asia is said to be particularly vulnerable due to rather weak environmental standards. The preliminary findings at RIS indicate that the developing countries' share of exports in the environmentally dirty products is increasing. It is argued that many dirty industries have already migrated from developed countries to this region (Bharucha, 1994). Mollerus (1995) examined the issue of migration of dirty industries with special reference to SAARC. Export data was examined to identify whether there have been significant changes of world trading patterns of products manufactured by polluting industries for the period 1982-92. Dirty industries in the SAARC countries are identified on the basis of pollution abatement and control expenditure of the US manufacturing industries. The results indicate that half of the polluting industries products exported from the SAARC countries represent a decline in world trade in which the products of the SAARC polluting industries are gaining in market share. This shows the significance of dirty industries in the SAARC region.

Tobey (1990) tested a similar hypothesis, this time across countries, using a Heckscher-Ohlin model of international trade. He chose 24 pollution-intensive industries (defined as those that in the US have abatement costs) at the three-digit SITC level in mining, primary metals, pulp and paper, and chemicals. Using an ordinal approach to measure regulatory stringency, countries' environmental stringency were ranked on scale of 1 to 7. Three types of regression analysis are used to look for the impact of environmental standards on international trade flows of these across countries, using data from 1970 and 1984. Tobey concluded that the effect of measure of stringency of environmental policy on changes in net exports is insignificant statistically. However in chemical industry in particular this relationship is positive. His results are confirmed by a similar analysis by Beers and van den Bergh (1997) for 1992.

Gallagher and Ackerman (2000) found no empirical data to support the pollution haven hypothesis. They developed a simple, partial equilibrium model to analyse the effects of a change in a trade policy. Their model analyses the relative concentrations of 'clean' and 'dirty' industries in two nations or regions, before and after the policy change.

Clark, Marchese and Zarrilli (2000) investigates whether the cost of environmental regulation influences the international location of polluting industries. Industries that operate production facilities in developing countries are identified through their use of the offshore assembly provisions in the US tariff code. Pollution intensity of industrial output is found to significantly reduce the probability of conducting offshore assembly in developing countries. This finding contradicts the argument that developing countries are becoming pollution havens as a result of offshore assembly.

Banerjee and Chattopadhyay (2000) did a study to check if trade liberalisation in India has encouraged our country to specialise in pollution intensive industries and see if the benefits bought by trade liberalisation exceed the cost of additional environmental degradation. They have done so by finding the ratio of pollution content of Indian exports to that of Indian imports at three different points of time: 1985, 1992 and 1996. The pollution content of the export or the import bundle is defined, as the total abatement cost implicit in the export or import bundle. They have estimated total pollution content of a commodity in terms of total abatement cost involved in the entire process. Though they are conceptually correct, the use of United States' pollution abatement could have lead to misleading results.

Impact of Standards and Capacity Building

In the existing literature it is generally acknowledged that developing countries have had to adjust their production processes in response to changing environmental regulations in the developed countries. Measures such as pesticide residue levels permitted in foodstuffs, emissions standards for machines, and packaging requirements have exerted pressure on the exporters. However, what remains to be seen is whether these measures have had any significant impact on trade. The emergence of global marketplace has resulted in debate on the possible impact of environmental standards and regulations on national trade competitiveness. In the developed countries, the major worry is that the introduction of more stringent environmental regulations in the domestic market would result in loss of competitiveness *vis-à-vis* industries in countries where standards are less stringent. Developing countries, on the other hand, are concerned that increasingly stringent regulations in developed countries could adversely affect market access for their export products. Kim (2000) has analysed how assistance to cope up with international environmental agreements may become more useful for the developing countries. Various studies indicate that the increasing environmental regulations have increased the cost of compliance faced by developing nations. As a result, the competitiveness of these countries, facing high compliance cost, is said to have reduced.

Most of the studies on competitiveness and environmental standards are based on ECC. Parikh *et al.* (1993) analysed the effects domestic environmental policies of foreign countries on India's exports focussing on two sectors: leather and shellfish. The results indicated that the increasingly stringent export standards have contributed to a rise in the cost of production, especially in the leather sector, where costs using the more environmentally friendly methods are nearly three times higher. Mahtab (1997) examines the prospect of leather industry in Bangladesh. He points out that the environmental costs arising from production of leather are not too enormous in Bangladesh and stresses the need of the Government and the industry to focus on environmental problem areas. The study by Murty and Prasad (1998) conducted a sample survey of factories belonging to water-polluting industries in India to analyse the environmental consequences of trade. It reveals that most of them have not made adequate expenditures to realise the standards and the expenditures needed to do so can heavily affect the present cost structure. Bharucha (2000) conducted a detailed questionnaire-based survey with number of industry representatives and institutions in tea, dyes, agriculture products and processed foods, marine products, leather, textiles, and the refrigeration and air- conditioning industry in India.

The main conclusion emanating from the studies is that compliance with external eco-standards often necessitated the import of inputs and technology, which were likely to raise the cost of production and price of output. Since competitiveness of many Indian exports is based on price factors, such price rise could hamper India's competitiveness. Jha and Vossenaar (1999) point out that the competitiveness effects of increased environmental compliance costs largely depend on the share of corresponding cost categories in total production costs. Thus, low valued products may be relatively vulnerable. For example, packaging requirements may have more significant effects on certain fruit and vegetables than on high-value added products. They also give the example of textile industry in India. They estimate that average raw material, of which dyes are a significant proportion, represent around 60 per cent of the cost of production. Thus, switching to more expensive environment-friendly dyes may have a significant bearing on the production costs.

Several studies have stressed the need for access to and transfer of environmentally sound technologies (ESTs). Jha and Teixeira (1998) discuss the number of constraints that stand in the way of transfer of EST. The paper points out that the single most important factor inhibiting the transfer of EST to developing countries is its cost, exacerbated by difficulties in obtaining credit and foreign exchange. Apart from direct costs of acquiring there are a number of indirect costs. There are also concerns about the viability of ESTs if applied to only the export sector.

The paper also identifies: difficulties that arise from attempting to directly apply EST designed in OECD countries to the conditions that prevail in the developing countries; the barriers posed by the standard setting in OECD countries that does not take sufficient account of needs of exporting firms in the developing countries; the cost of patents and of purchasing or licensing EST; and the difficulties associated with foreign direct investment. According to the paper, OECD countries prefer technology transfer on commercial terms with some financial assistance, while the developing countries claim that the transfer of EST should be made on a preferential and concessional basis.

Achanta *et al.*, (1997) discuss the transfer of EST with special reference to India. They point out the importance of transfer of EST to Indian industry and need for institutional and policy changes needed to accelerate the transfer. Saqib and Kaushik (2001) discuss the scope of exporting environmentally friendly products for organic farming in India. Though organic farming has a tremendous scope in India, there exist a number of bottlenecks for farmers who are interested in organic farming and for the industry which wants to process and export these products. These impediments are in the area of production, marketing and infrastructure. Certification is seen as a barrier to the small growers due to its costs. Standards are too high and are creating unfair barrier to production and trade. India has no local certification systems for organic products and farmers have to depend on foreign certification like IFOAM and SKAL. This is very expensive and is feasible for big holdings only.

The Advanced Technology Assessment System (ATAS) in 2000 examines the role publicly funded research and publicly owned technologies in transfer and diffusion of ESTs, with particular attention to developing countries. In the volume, Tripathy (2000) examined this issue in case of India. He points that the privately funded R&D organisations may not like to part with ESTs. Hence, in the coming decades publically funded ESTs will be the only source of affordable and accessible technology to most firms that are not capable of importing appropriate ESTs. The study found that there are many ESTs in the public sector domain. In certain industry sectors, such as the pollution prone small-scale sector, ESTs are spreading fast because of State intervention. But the spread is limited in the heavy industries sector, where large capital inputs may be required. The informal sector poses a daunting problem as no policies are in place to spread ESTs and policy focus is required. The paper also discusses the financial aspects of transfer of ESTs. It points out that external funding has been instrumental in transferring ESTs from the developed to selected industrial sectors. Some of these transfers have focussed on ESTs appropriate to those segments of the economy covered by multilateral agreements such as the Montreal Protocol and agreements to reduce the emissions of gases with global warming potential. The Government of India also provides special assistance, including finance on concessional terms, for activities that contribute directly to improvement of technology. Khan (1997) studies the scope for technical assistance for production of environmentally friendly production in Pakistan.

Concluding Remarks

The effect of environmental measures on trade has become a matter of concern for the proponents of trade liberalisation. As the liberalisation has taken place, there has been a global proliferation of technical measures, particularly in developed nations. There has also been a rise in the number of studies that recognise these technical measures as a barrier to trade, either implicitly or explicitly. The literature has shown that it is difficult for developing countries to trade with the developed countries, due to differences in quality requirements.

The proliferation of environmental and other technical standards is accompanied by rise in trade of environmentally sensitive goods. A number of studies indicate that environmental standards of North have affected the trade prospects in certain important export oriented sectors of developing countries.

As far as migration of pollution intensive industries is concerned, existing literature fails to throw much light. Although there seems to be plenty of anecdotal evidence that policy makers and industrialists take industrial flight seriously, there is only a limited amount of empirical evidence that industrial flight exists.

The debate about competitiveness effects of environmental regulations has focussed on the costs of compliance. While the compliance costs at this level appear higher in a static analysis, dynamic analysis shows lower costs since cleaner technologies may result in cost savings over the long term. Gradual introduction, exceptions, and rebates mitigate the effects of international competitiveness and trade. Complete or partial financing of research into ESTs and direct cash grants to developing countries may contribute to the dynamic adjustment of firms in developing countries to higher standards.

References

Abimanyu, A. (1996). "Impact of Free Trade on Industrial Pollution. Do Pollution-Havens Exist?", *ASEAN Economic Bulletin*, 13, pp. 39-51.

Achanta, A.P. Dadhich, P. Ghosh and L. Noronha (1997). "The Transfer of Environmentally Sound Technologies (EST) with Special Reference to India", in V. Jha, G. Hewison and M. Underhill (eds.), *Trade, Environment and Sustainable Development: A South Asian Perspective*, Macmillan Press, London, pp. 185-202.

Andersson *et al.* (1995). *Trading with the Environment, Ecology, Economics, Institutions, and Policy*, Earthscan Publications, London.

Banerjee, S. and S. Chattopadhyay (2000). "Globalisation and Pollution Trading: The Indian Experience", *Calcutta University Seminar Paper #1/2000-01*.

Beers, C. van and C.J.M., van den Bergh (1997). "An Empirical Multi-Country Analysis of the Impact of Environmental Regulation on Trade Flows", *50 Kyklos*, fasc. 1 at 25.

Bhagwati, J. and R. Hudec. (1996). *Fair Trade and Harmonisation-Pre-requisites for Free Trade?*, The MIT Press.

Bharucha, V. (1994). *Impact of Environmental Standards and Regulations on India's Exports*, study by UNCTAD.

————. (2000). "The Impact of Environmental Standards and Regulations Set in Foreign Markets on India's Exports", in Veena Jha, G. Hewison and M. Undenhills (eds.) *Trade, Environment and Sustainable Development: A South Asia Perspective*, Macmillan Press, London, pp. 123-42.

Cato, J.C. (1998). *Economic Issues Associated with Seafood Safety and Implementation of Seafood HACCP Programmes*. FAO, Rome.

Chang, Seung Wha (1997). "GATTing a Green Trade Barrier: Eco-Labelling and the WTO Agreement on Technical Barriers to Trade", *Journal of World Trade*, 31 (1), pp. 137-159, February.

Chaturvedi, S. and G. Nagpal (2001). "Product Standards and Trade in Environmentally Sensitive Goods: A Study of South Asian Experience", *RIS Discussion Paper #22*.

Clark, D.P., S. Marchese and S. Zarrilli (2000). "Do Dirty Industries Conduct Offshore Assembly in Developing Countries?", *International Economic Journal*, 14(3), August.

CUTS (1995). *Environmental Conditions in International Trade*, CUTS, Jaipur.

Eskeland, Gunnar and Ann Harrison (1998). "Moving to Greener Pastures: Multinationals and the Pollution-Haven Hypothesis", *World Bank Policy Research Paper* No. 1744. World Bank, Washington D.C.

Esty C.D. and Damien Geradin (1998). "Environmental Protection and International Competitiveness: A Conceptual Framework", *Journal of World Trade*, 32(3), pp.5-46.

Finger, J.M., and P. Schuler (1999) "Implementation of Uruguay Round Commitments: The Development Challenge", paper presented at the conference *Developing Countries in a Millennium Round*, Geneva, September.

Gallagher, K. and F. Ackerman (2000). "Trade Liberalisation and Pollution Intensive Industry in Developing Countries", *Working Paper*, 2000-03, *Global Development and Environment Institute*.

Grossman, G. and A.B. Krueger (1992). "Environmental Impacts of a North American Free Trade Agreement", *CEPR Discussion Paper No*. 644, Centre for Economic Policy Research, London.

Han, Ki-Ju and John B. Braden (1996). *Environment and Trade: New Evidence from U.S. Manufacturing*, University of Illinois, Department of Economics, *mimeo*.

Hufbauer, G., B. Kotschwar, J. Wilson (2000). "Trade, Standards and Development: Perspectives for Central America", paper presented at *Workshop on Trade Facilitation, Regulation, and Standards: The Development Challenge in Central America* held in Panama City, June 27-29.

Intal, P., E. Medalla, M. Los Angeles, D. Israel, V. Pineda, P. Quintos and E. Tan (1994). *Trade and Environment Linkages: The Case of the Philippines*, Philippines, Manila Institute of Development Studies.

Jha, Veena, Anil Markandya, Rene Vossenaar (1999). *Reconciling Trade and the Environment: Lessons from Case Studies in Developing Countries*, Edward Elgar, UK.

Jha, Veena and Ana Paola Teixeira (1998). "Trade, Environment and the Transfer of Environmentally Sound Technology", in Jha, Hewison and Underhill (eds.), *Trade, Environment and Sustainable Development: A South Asian Perspective*, Macmillan Press Ltd., London.

Jha, Veena and Rene Vossenaar (1999). "Environmental Policy, Market Access and Competitiveness: The Experience of Developing Countries", in Veena Jha, Anil Markandya and Rene Vossenaar (eds.), *Reconciling Trade and the Environment: Lessons from Case Studies in Developing Countries*, Edward Elgar Publishing Limited, UK.

Kalt, J.P. (1998). "The Impact of Domestic Environmental Regulatory Policies on US International Competitiveness, in A.M. Spence and H.A. Hazard, *International Competitiveness*, (eds.), Harper and Row, Ballinger, Cambridge, M.A., pp. 221-62.

Khan, Akhtar Hasan (1997). "Making Trade and Environmental Policy Making Mutually Compatible in Pakistan", in Jha, Hewison and Underhill (eds.), *Trade, Environment and Sustainable Development: A South Asian Perspective*, Macmillan Press Ltd., London.

Kim Hyung-Jim (2000). "Subsidy, Polluter Pays Principle and Financial Assistance among Countries", *Journal of World Trade*, 34(6), pp. 115-141.

Leebron, David W. (1996). "Claims for Harmonisation: A Theoretical Framework", *Canadian Business Law Journal*, pp. 63-106.

Leonard, H. J. and C. J. Duerksen (1980). "Environmental Regulations and the Location of Industry: An International Perspective", *Columbia Journal of World Busines*, summer, pp. 52-68.

Letchumanan, R. (1998). "Trade, Environment and Competitiveness: Testing the 'Pollution-Haven' Hypothesis from Technology Perceptive", *UNU Working Paper* No. 43.

Low, P. and A., Yeates (1992). "Do 'Dirty' Industries Migrate?" in P., Low. (ed.), *World Bank Discussion Paper*, No. 159, *International Trade and the Environment*, pp-89-104.

Lucas, R.E.B., D. Wheeler, and H. Hettige (1992). "Economic Development, Environmental Regulation and the International Migration of Toxic Industrial Pollution: 1960-1988", in P. Low (ed.), *International Trade and the Environment. World Bank Discussion Paper* 159, Washington D. C.

Mani, M. and D., Wheeler (1997). "In Search of Pollution-Havens? Dirty Industry in the World Economy, 1960-1995", *Journal of Environment and Development*, 7(3), pp. 215-247.

Markandya, Anil (1999). "Overview and Lessons Learnt" in Veena Jha, Anil Markandya and R. Vossenaar (*et al.*) *Reconciling Trade and the Environment: Lessons from Case Studies in Developing Countries*, Edward Elgar Publishing, UK.

Mahtab, Fasih Uddin (1997). "The Sustainable Development of Leather Industries in Bangladesh", in Jha, Hewison and Underhill (eds.), *Trade, Environment and Sustainable Development: A South Asian Perspective*, Macmillan Press Ltd., London.

Maskus, K., J., Wilson and T., Otsuki (2000). *Quantifying the Impact of Technical Barriers of Trade-A Framework of Analysis.*

Mollerus, Roland (1995). *Environmental Standards: Impact of SELAs Competitiveness and Market Access. Trade and Environment: International Debate*, UNCTAD/SELA.

Murphy, M.P., and A., Shleifer (1997). "Quality and Trade", *Journal of Development Economics*, 53, pp.1-15.

Murty, M.N. and U.R. Prasad (1998). "Environment and Trade: Developing Country's Perspective and Indian Experiences", *Working Paper Series IEG No. E/192/98*, February.

Mutasa, M.P. and T., Nyamandi (1998). "Report of the Survey on the Indentification of Food Regulations and Standards within the Africa Region Codex Member Countries that Impede Food Trade", paper was presented at workshop on *Codex and Harmonisation of Food Regulations*, Harare.

Otsuki, T., J.S., Wilson and M., Sewadeh *(et al.)* (2000). *A Race to the Top? A Case Study of Food Safety Standards and African Exports*, Development Research Group (DECRG), World Bank.

OECD (1991). *The OECD Environment Industry: Situation, Prospects and Government Policies*, Paris

————. (1993). *Environmental Policies & Economic Competitiveness*, Paris.

Parikh, J.K., V.K. Sharma, U. Ghosh and M.K. Panda (1994). *Trade and Environment Linkages: A Case Study of India*, report prepared for United Nations Conference on Trade and Development, August.

Pearson, Charles (ed.) (1987). *Multinational Corporations, Environment, and the Third World*, Duke University Press, Durham, NC, pp.3-31.

————. (ed.) (2000). *Multinational Corporations, Environment, and the Third World*, Durham, Duke University Press, NC.

Robertson, D. and A., Kellow (2000). *Globalisation and the Environment: Risk Assessment and the WTO*, Edward Elgar.

Rutgeerts, A. (1999). "Trade and Environment: Reconciling the Montreal Protocol and the GATT", *Journal for World Trade* 33 (4), pp. 61-86

Saqib, M. (2000). "Technical Barriers to Trade and the Role of Indian Standard Institutions", *RGICS Working Paper Series No. 19.*

Saqib, M. and A., Kaushik (2001). "Exporting Envrionment Friendly Products: Scope for Organic Farming in India", *RGICS Working Paper Series No. 21.*

Sorsa, P. (1992). "GATT and Environment: Basic Issues and Some Developing Country Concerns", in P. Low (ed.), *International Trade and the Environment, World Bank Discussion Paper No. 159*, Washington, DC.

————. (1994). "Competitiveness and Environmental Standards: Some Exploratory Results", *Policy Research Working Paper 1249*, The World Bank, Washington, DC.

Tieje Christan (1995). "Voluntary Eco-labelling Programmes and Questions of State Responsibility in the WTO/GATT Legal System", *Journal of World Trade*, Vol. 29(5), pp. 123-158.

Tobey, James (1990). "The Effects of Domestic Environmental Policies on Pattern of Trade: An Empirical Test", *Kyklos*, 43, pp.191-209.

Tripathy, Upendra (2000). "The Role of Publicly Funded Research in the Transfer and Diffusion of Environmentally Sound Technologies: Case of India", in the *Role of Publicly Funded Research and Publicly Owned Technologies in the Transfer and Diffusion of Environment Sound Technologies*. UNCTAD, ATAS XII.

Wyatt, Christine (1997). "Environmental Policy Making, Eco-Labelling and Eco-Packaging in Germany and its Impact on Developing Countries", in Jha, Hewison and Underhill (eds.), *Trade, Environment and Sustainable Development: A South Asian Perspective*, Macmillan Press Ltd., London.

Xing, Y. and Kolstad, C.D. (1996). "Do Lax Environmental Regulations Attract Foreign Investment?" *National Bureau of Economic Research Workshop on Public Policy and the Environment*, Cambridge, MA.

Xinpeng Xu (1999). "Do Stringent Environmental Regulaton Reduce the International Competitiveness of Environmentally Sensitive Goods? A Global Perspective", *World Development*, 27(7), pp. 1215-1226.

Part II

Experiences from South Asia

4

Environmental Related Barriers in the EU

S.K. MOHANTY
T.R. MANOHARAN

Introduction

A major issue relating to trade and environment is the manner in which developed countries use 'environment' as a protectionist measure in international trade. Such protectionist measures usually called "Green Protectionism" has many dimensions in trade (Esty, 1994). It is argued that while tariffs have declined as a result of liberalisation process, the environmentally related non-tariff measures (NTMs) and other technical standards have emerged as significant factors in determining world trade.

There are three important consequences of 'green protectionism' on trade prospects of developing countries. First, many of their export commodities will face entry barriers to important markets in the form of higher environmental regulations, process and production methods (PPMs) or certification schemes. Second, adjusting to higher environmental standards will make their economies less competitive in overseas markets since the ability of developing countries to implement environmental standards set by foreign markets are limited either by financial and technical constraints or by the size of the domestic demand for environmentally friendly products. Third, differences in environmental standards will cause relocation of 'dirty industries' from developed (with higher environmental standards) to developing (with lower environmental standards) countries.

The United Nations Conference on Environment and Development (UNCED)—Rio Earth Summit—held in 1992 fully recognised the role of trade in economic development but cautioned about the damaging effect of trade due to the unsustainable utilisation of natural resources and environment (UNCED, 1992). The Earth Summit influenced the process of trade liberalisation during the post-Rio period. The establishment of Committee on Trade and Environment (CTE) is an important step taken in the Ministerial Meeting of UR held at Marrakesh in 1994. When WTO was established in 1995, both environment and trade communities were very much keen to know the work programme and activities of CTE. The committee held discussions on its identified 10 items on trade and environment but could not come up with any amicable solution so far. The CTE noticed that the issues on trade and environment are complex. Subsequently, the discussions at CTE were based on 'cluster approach' under the two themes of (i) market access and (ii) the linkages between the multilateral environment and trade agendas (WTO, 1999). India and many other developing countries opposed any attempt to include non-trade issues such as environment and labour in the negotiating framework. However, they had to compromise this with the EU and other developed countries in the fourth Ministerial Meeting of WTO held at Doha in 2001 when the decision was taken to begin negotiations on selected issues on trade and environment.

Market access is an important issue of the developing countries. The environmental standards in developed countries are very high. Further, many of them have increased their environmental standards and other measures in the recent past. The higher environmental standards in developed countries are not found to have adversely affected their trade. But it is a matter of concern for the developing countries especially for small and medium scale enterprises (SMEs) in those countries. The problems include additional costs and lack of environmentally friendly technology in complying with such standards.

Of various studies available on trade and environment, a few of them have addressed the issue of environmentally related non-tariff barriers in developing countries' trade (Dean *et al.*, 1992; Jaffee *et al.*, 1995; Manoharan *et al.*, 2000). Sorsa (1994), while analysing environmentally sensitive products, found that between 1970 and 1990, the developing countries' share in the imports of environmentally sensitive commodities (ESCs) has declined in the selected European countries. UNCTAD initiated a number of case studies in the developing countries to understand the linkages between trade and environment. Such studies pointed out that environmental standards adversely affected many export oriented sectors such as textiles, chemicals, leather and marine products. The SMEs face more difficulties as compared with big companies and multinational corporations (MNCs) (Jha, 1997).

NTMs in the European Union and South Asia's Trade: Some Key Questions

The formation of the EU in the early nineties created high hopes among developing countries, including South Asian countries, that the new regional trading arrangement (RTA) would improve market access of extra-regional countries in view of the expanding market of the Union. Against such expectations, the net increase in the levels of trade of the South Asian countries with the EU has been dismal. It is the feeling of the regional countries of the South Asia that the persistence of peak tariffs, specific tariffs, tariff escalation and NTMs have significantly eroded the prospects of the South Asian exports to the EU. Besides these measures, the environmentally sensitive NTMs have, to a large extent, constrained the market access of South Asian countries in the EU. The complexities involved in both trade and non-trade barriers raise a number of issues for discussion in the context of trade prospects of South Asian countries in the EU. The possible concerns in this regard are:

- Is the tariff structure of the EU invariably low across the product groups? Or does the average level of tariff in different product groups vary significantly from one sector to another, thus, allowing for discrimination in the level of protection between sectors?
- To what extent are the imports of the Union protected by NTMs? And what is the role played by environment related NTMs in affecting export prospects of South Asian exports in the Union?
- What is the coverage of specific tariff in the customs tariff of the EU?
- Is it the case that the specific tariffs combined with NTMs have downsized the level of market access of extra-regional trading partners including South Asian countries in the EU?
- What are the prospects of improving South Asian exports to the EU by giving priority to regional countries over its

competitors in the trade of those commodities, which are subject to environment-related NTMs?

The important issues faced by the exporters of developing countries on market access, as reported by UNCTAD are: lack of timely and accurate information, lack of scientific data for specific thresholds or limit values, inconsistent application of procedures, uncertainty due to rapidly changing environmental and sanitary requirements in overseas markets, varying standards and regulations in different markets and the costs and difficulties of testing and verification procedures.

The main objectives of the present study are the following:

(i) To examine the coverage of the environment-related NTMs in different product groups of the EU's imports, and the role of other NTMs in limiting the market access of developing countries.

(ii) To analyse the structure of *ad valorem* tariff and specific tariff in different product categories prevalent in the EU.

(iii) To examine the linkages between TBs and NTMs in protecting the EU market and its implication on the export prospects of South Asian countries.

Environment Related NTMs

Developed countries have imposed a number of environment related NTMs to regulate trade. An analysis of such NTMs is a difficult task because the available data sources such as TRAINS database of UNCTAD, do not provide information on environment related NTMs as a distinct category. The list of NTMs provided by TRAINS is too aggregate in many countries, especially in the US where only four NTMs are provided in the list. This aggravates the difficulty in identifying and analysing environment related NTMs. Therefore, we have restricted our study only to the EU and carried out the analysis on the basis of available information from TRAINS database.

In this study we have selected the EU because of two reasons. First, the EU market is a major and important market for South Asia. Second, the availability of database on some of the environment related NTMs, is better in the case of the EU.

Table 4.1

Selected Non-tariff Measures Applied by the European Union

Reference Code (As per TRAINS) (3 Digit)	*Non-tariff Measures*
	(A) Environment Related
110	Prior Authorisation (CITES)
113	Quota in Relation to Montreal Protocol
114	Prohibition for Environmental Protection
116	Product Charct. Requirements (Health)
	(B) Others
101	Tariff Quota
102	Seasonal Tariff Rates
103	Variable Charges n.e.s
104	Anti-dumping Investigations
105	Anti-dumping Duties
106	Automatic Licensing Requirements
107	Import Monitoring
108	Retrospective Surveillance
109	Non-automatic Licence
111	Quotas
112	Bilateral Quota
115	Import Monopoly

Note: The classification 'Environment Related' is made for the purpose of this study.
Source: TRAINS 6.0, UNCTAD.

Of the available information on 16 NTMs in the EU, 3 are seen to be imposed exclusively for environment purposes. They are: (i) prior authorisation as per Convention on International Trade in Endangered Species of Wild Fauna and Flora (CITES) (code:110), (ii) quota in relation to Montreal Protocol (code:113) and (iii) prohibition for environmental protection (code: 114). Therefore, these 3 are taken as environment NTMs in the present study. NTM on product characteristic requirements for health (116) is also taken into consideration because health and environment are closely interrelated. A description of these NTMs is given in the Table 4.1.

The first two environmental NTMs are based on Multilateral Environmental Agreements (MEA) *viz.*, CITES and the Montreal Protocol.

CITES

The Convention on International Trade in Endangered Species of Wild Fauna and Flora (CITES) was concluded in 1973 and came into force on 1 July 1975. Having a current membership of 132 countries, it established the legal framework for regulating trade in endangered and other species globally.

The four Appendices to the convention form the basis for implementation of the treaty. The first three Appendices list species for different levels of trade restriction on each category. Appendix IV makes provision for the issue of permits for international trade in the species listed in the three Appendices. The Appendices are amended at least once every two years, in accordance with procedures and criteria set forth in the treaty and through resolutions adopted by the Conference of the Parties (COP) at their biennial meetings.

Appendix I shall include all species threatened with extraction that are or may be affected by trade. Trade in specimens of these species must be subject to particularly strict regulation in order to avoid further endangering of their survival and must only be authorised in exceptional circumstances. Appendix I contained about 450 species, majority being well known endangered animals such as elephant, lion, tiger, cheetah, leopard and sloath bear.

Commercial trade in species listed in Appendix I is prohibited and non- commercial trade is allowed only in exceptional circumstances provided it will not damage their chances for survival. For all transactions in the Appendix I species, an export permit from the country where the wildlife originates or re-export certificate from the re-exporting country, and an import permit from the recipient country are required. The import permit must be issued before any transaction involving the Appendix I species can occur.

The import permit can be granted only if (a) the importation will not be for purposes detrimental to the specie's survival, (b) the importation will not be primarily commercial and (c) if live wildlife is involved, the specimen will be assured of a suitable home.

The export permit can be issued only if (a) the wildlife was obtained legally, (b) the wildlife will not be harmed during shipping and (c) an import permit has already been granted.

Appendix II shall include all species that although not necessarily now threatened with extinction may become so unless trade in specimens of such species is subject to strict regulation in order to avoid utilisation incompatible with their survival; and other species that must be subject to regulation in order that trade in specimens of certain species referred to in the above sub-paragraph may be brought under effective control.

CITES permits commercial trade in Appendix II species only if the country of origin has issued an export permit or a re-export permit unless the proper government agency has certified that the export will not be detrimental to the species survival. Import permits are not required. Appendix II is much larger than Appendix I and contains more than 2300 animals' species and over 24,000 plant species. The larger size of Appendix II is because the entire families of species are listed in it whereas in Appendix I individual species are listed.

Appendix III shall include all species that any party identified as being subject to regulation within its jurisdiction for the purpose of preventing or restricting exploitation, and as needing the co-operation of other parties in the control of trade.

Thus, CITES gives parties the option of listing under this Appendix native species that are already protected within their own borders. Operation in enforcing their own wildlife trade regulations. So if a country lists certain species under Appendix III, and not under any other Appendices it then implies that trade in that particular species is regulated in that country, and import of that species or its parts from that country is illegal without a CITES export permit issued by the management authority of that country. Trade in Appendix III species, thus, requires an export permit from the listing country or a certificate of origin when the specimen is exported from other countries. The parties shall not allow trade in specimens included in the three Appendices except in accordance with the provisions of the Convention.

Montreal Protocol

On 16th September 1987, The Montreal Protocol on substances that deplete the ozone layer was finally agreed upon at the headquarters of the International Civil Aviation Organisation (ICAO) in Montreal, and it came into force on January 1, 1989. Currently, 172 countries have signed the Protocol.

The basic objective of the Protocol is to protect ozone layer through control of substances that deplete ozone layer in a manner that is likely to result in adverse effects on human health and the environment. The measures taken for this purpose should be based on relevant scientific knowledge, taking into account technical and economic considerations.

In this regard, the developmental needs of the developing countries is also taken care of. Article 2 of the Protocol describes the control measures to be taken against global emissions of selected chloroflurocarbons (CFCs) and other substances such as halons, carbontetrachloride, trichloroethance, hydrochlorofluro-carbons and methyl bromie listed in Annexes A, B, C and E. Annexure A provides 8 substances of which 5 are in Group I and the remaining in Group II. The ozone depleting potential is also given in the Annexures. The ozone depleting potentials are estimates based on existing knowledge and will be reviewed and revised periodically. Annexure B contains a total of 12 substances in three groups of which 10 are in Group I and the remaining 2 are distributed 1 each in the other two groups. Annexure C contains 74 substances of which 40 are listed in Group I and the remaining in Group II. Annexure E has one substance which is methyl bromide having ozone depleting potential of 0.6.

Annexure D provides a list of products containing controlled substances specified in Annexure A. About six product categories are given in this Annexure. They are:

1. automobile and truck air conditioning units (whether incorporated in vehicles or not;
2. domestic and commercial refrigeration and air-conditioning/ heat pump equipments (Example, refrigerators, freezers, dehumidifiers, water coolers, ice machines, air conditioning and heat pump units, etc.);
3. aerosol products except medical aerosols;
4. portable fire extinguisher;
5. insulation boards, panels and pipe covers; and
6. pre-polymers.

In the case of CFCs, Article 2A1 says: "Each part shall ensure that for the twelve month period commencing on the first day of the seventh month following the date of entry into force of this Protocol and in each twelve month period thereafter, its calculated level of consumption of the controlled substances in Group I of Annexure A does not exceed its calculated level of consumption in 1986. By the end of the same period each party producing one or more of these substances shall ensure that its calculated level of production of the substances does not exceed its calculated level of production in 1986, except that such level may have increased by no more than ten per cent based on the 1986 level." For developing countries "such level may have increased by no more than 10 per cent based on the 1986 level."

The second, fourth, seventh and ninth meetings of the parties of Montreal Protocol took more stringent measures on emission levels. Para 4 of Article 2A says, "Each party shall ensure that for the 12 month period commencing on 1 January 1996, and in each 12 month period level of consumption thereafter, its calculated level of consumption of the controlled substances in Group I of Annex A does not exceed zero. Each party producing one or more substances shall, for the same periods, ensure that its calculated level of production of the substances does not exceed zero." For developing countries, considering their domestic needs, the calculated level of production may exceed that limit by up to 15 per cent of its calculated level of production in 1986.

The non-tariff measures applied by the EU on the basis of the Montreal Protocol will be covered for more products in the coming years. Though developing countries have been given special provision in the Protocol, the effect of the Protocol on trade prospects of developing countries needs to be fully addressed.

While the NTMs on the basis of Montreal Protocol and CITES are imposed as a result of MEAs, a number of restrictions and prohibitions are made on several important product categories on basis of environmental grounds. This includes restrictions on the use of certain chemicals, packaging requirements, etc.

Prohibition for Environmental Protection

The EU imposed import prohibition on a variety of products for environment protection. Imports of ozone depleting substances such as CFCs and halons from countries which are not parties to Montreal Protocol are prohibited. A number of chemicals and other substances which affect public health and environment are prohibited. This includes Poly Chlorinated Biphenyle (PCBs) and Poly Chlorinated Tetraphenyle (PCTs) (concentration exceeding 0.005 per cent of weight), Benzidine, 4-nytrobiphenyl, 4-aminobiophenyl, paints which contain neutral anhydrous carbonate, lead hydrocarbonate and lead sulphates, certain antifouling paints containing mercury compounds, arsenic compounds, and organostannic compounds and wood preservatives.

The EU prohibited the use of cadmium and its compounds as pigment or stabliser in the products such as packaging material, fittings for furniture, clothing and accessories, etc. Similarly, the marketing and use of all plant protection products is prohibited. Imports of skins of certain animals such as beaver, otter, wolf, etc. and the products derived from those skins are also prohibited.

Product Characteristic Requirement for Health

The EU has taken various measures for health and sanitary conditions for the imported products. For instance, imports of fresh, chilled, frozen meat/fishery products to the EU should meet certain health and hygiene standards. These imports should come from the EU-approved establishments (premises where products are prepared, processed, chilled, frozen, packaged or stored). The imports should bear an animal health certificate and public health certificate drawn by an EU-approved official veterinary authority of the exporting country. Meat products should bear a 'Health Mark'. Imports of these products as well as of fresh fruits and vegetables are subject to inspection upon arrival into the EU. A number of health and quality conditions are imposed on food products and non-food products having direct influence on human health. A good number of products such as soap, cosmetics, toilet preparations, fertilisers, pesticides, detergents, medicines, plastic packaging for food, etc. are subject to these requirements.

Methodology

In view of the broad objectives of the study we have used TRAINS database for the empirical analysis. The study requires data pertaining to different types of NTMs (separate for environment-related, trade related and health-related), MFN *ad valorem* tariffs and specific tariffs in the EU.

For the present study, we have used trade, *ad valorem* tariff, specific tariff and NTMs of the EU for examining the relationship between these variables in providing effective protection to the domestic sector of the Union. While NTMs, *ad valorem* tariff and specific tariffs are presented at the national lines, bilateral trade data are given at the HS sub-heading level. The level of desegregation associated with national lines varies from one country to another. The EU defines its national lines at 8-and 10-digit level. Combining both 8- and 10-digit HS level products, the EU has as many as 14251 national lines. While most of the products are defined at 8-digit HS level, some of them are defined at more desegregated level of 10-digit level. However, bilateral exports of the EU are given at 6-digit level. In 1997, the EU imported as many as 5106 products at 6-digit level from the global economy.

It is evident from the literature that the level of NTMs increased by many-folds, particularly in the post-Uruguay Round period in comparison with the pre-negotiation period. We are unable to examine the position of pre-and post-Uruguay situations due to lack of information regarding NTMs during the pre-URTN period. Therefore, our analysis is mostly restricted to cross-section analysis, based on most recent year data. We have used TRAINS, winter 1998, UNCTAD, CD for the present study. While the latest trade data is available for 1997, NTMs, *ad valorem* tariff and specific tariffs are given for the year July 1998. Despite the data constraints, we have taken the latest data available for this study. The individual member countries in the EU use several NTMs and many of them are neither consistent regionally nor compatible with the WTO framework.[1]

It is an intricate task to combine trade, NTMs and customs tariff in one place because the reported data are not provided by the UNCTAD or any other Multilateral Agencies at a given level of aggregation. While data on TBs and NTMs are available at national lines, trade data are in HS 'sub-heading level'. Moreover, aggregation of custom tariff rates may be unrealistic for negotiation purpose and

1. Such measures have adversely affected trade prospects of South Asian countries. For example, Germany required for the use of a specific type of dye in carpet for import, and this unilateral policy had adversely affected export prospects of some of the South Asian countries. We are, in fact, looking for such type of NTMs for the study. Due to lack of information on specific NTMs, we are restricting our analysis only to the available information.

the underlined danger of loss of information. However, we have tried to combine trade, tariff (both *ad valorem* and specific) and NTMs in a single spreadsheet. Moreover, attention has been paid to minimise loss of information. The purpose of this exercise is to analyse the extent to which environment-related NTMs affect export prospects of South Asia in the markets.

In the first place we have attempted to map EU's tariff, specific tariff and NTMs at the most desegregated product level (both at 8-and 10-digit level), since these variables are defined at the national lines. It is found that many products at the national lines are subject to either single or multiple NTMs.

The TRAINS[2] database provides information on 16 types of NTMs, for the EU. From the total of 16,[3] 3 of them are environment-related, one is health-related and remaining 12 are trade related NTMs. All three environment-related NTMs are WTO compatible. They are: a) prohibited under CITES (code 110); b) prior authorisation required under Montreal Protocol (code 113) and; c) prohibited due to environmental reasons (code 114). The number of products falling under CITES is relatively much larger than those of other two types of environment-related NTMs.

We have analysed information provided in the TRAINS database in two phases. In the first phase, we have mapped *ad valorem* tariff, specific tariff and NTMs for each national line separately. Occurrences of both environment-related, health-related and trade-related NTMs across the product lines are mapped separately. For these groups of NTMs, frequency[4] ratios are estimated separately.

In order to converge TBs and NTMs data with the trade data, we have taken aggregated figures of the NTMs and tariff data at the 6-digit level. We have estimated frequency ratios[5] and average tariff for

2. We refer to Winter 1998 version of TRAINS, UNCTAD.
3. A detailed discussion on the nature of NTMs in the EU is given in Section 2.1.
4. The frequency of NTMs of a HS product line indicates number of times different NTMs are against a product invoked by the importing country. For example, against the national line 'bromochloromethane' (HS 2903498010), the EU imposes two environment related NTMs namely, Montreal Protocol and prohibition in 1996. Therefore, the frequency of the product line is equal to two. For other product lines, frequency lines are estimated in a similar manner.
5. Frequency ratio of a broad product (at 6-digit level) is the ratio of number of micro-products (at national level) subject to one or more NTMs to total number of micro-product lines (at national level) falling under the same broad product (at 6-digit level).

each product line at 6-digit HS level. Very often, we find difficulty in computing average tariff for each product at 'sub-heading' level because of 'specific tariffs'. We have also estimated frequency ratios separately for environment-related, health-related and trade-related NTMs.

Results

In the post-Uruguay Round of trade negotiations, the agricultural sector continued to be one of the most protected areas of trade[6] in developed countries, and this has been more so in the EU. The protection of the agricultural sector is not only significant in terms of coverage of products but also in terms of intensity[7] of protection. The incident of NTMs is relatively much higher in the agricultural sector than that of the industrial sector. It is observed that both environment-related and health-related NTMs are more predominantly present in the primary sector as compared with the manufacturing sector. About half of the total number of agricultural products and one tenth of total manufacturing products are subject to one or the other forms of NTMs as shown in Figure 4.1. While 1910 product lines out of 4043 agricultural products are subject to single or multiple NTMs only 1016 products out of 10208 manufacturing products are subject to NTMs.

The extent to which different product groups are affected by the NTMs is examined by the frequency ratio. Table 4.2 presents number of national lines subject to environmental, trade and health related NTMs and their frequently ratios in the EU. It is interesting to note that both environmental and trade related NTMs are simultaneously invoked in both agricultural and industrial sectors, but the health related NTMs are imposed only on in the agricultural sector. In terms of coverage of products, both environment-related NTMs and health

6. Protection is referred here to denial of market access through different forms of NTMs, but not to customs tariff.

7. When a specific product is covered by one type of NTMs, the import of the product is restricted to some extent by the importing country. But when the same product is subject to several types of NTMs, the possibility of import of the same product is severely affected. In other words, market access of the specific product is denied by multiple conditions attached in the form of one or another NTM. Therefore, intensity of NTMs increases when more number of NTMs are invoked for one line of product. The possibility of having market access declines when intensity of NTMs increases.

Figure 4.1

Coverage of NTMs in the Imports of the EU

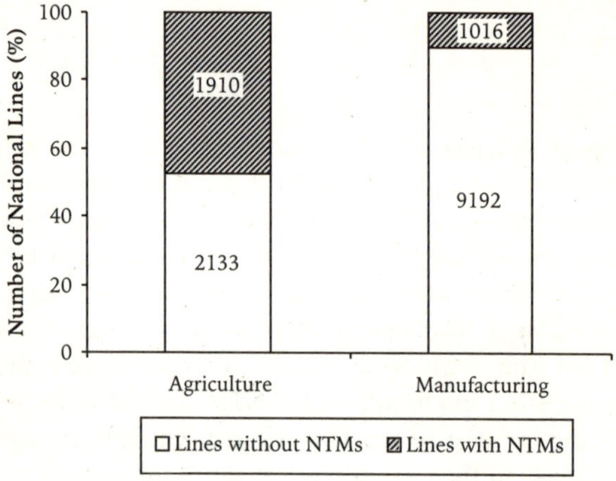

related NTMs are mostly concentrated in the agricultural sector. The number of products under each of these broad group of NTM is quite substantial in comparison with total number of agricultural products imported. Total number of products covered by environment related and health related NTMs constitutes almost two thirds of total number of products subject to trade related NTMs in the agricultural sector. The situation in the manufacturing sector is different from that of agricultural sector. The number of lines subject to the environmental related NTMs constitutes almost one third of that of trade related NTMs in the manufacturing sector.

In the literature, frequency ratio is very often used as an index for measuring the extent of protection conferred on import through NTMs. The effectiveness of a specific NTM, particularly in terms of import restraining capacity, may be difficult to ascertain. It may vary from one type of NTM to another in an importing country.

Therefore, ordering of NTMs on the basis of their effectiveness in a country is generally not seen in the literature. In the absence of quality-related NTMs analysis, frequency ratio is very often used to examine the coverage of the products in different sectors.

It is estimated that the overall frequency ratio of the EU is 20.5 per cent in 1998, thereby indicating that one fifth of the EU's National lines (for imports) is subject to one or multiple NTMs. The results show that the level of frequency ratio varies (overall frequency ratio and environment related frequency ratio) significantly from one sector to another (Figure 4.2.)

Figure 4.2

Composition of Frequency Ratios in the EU

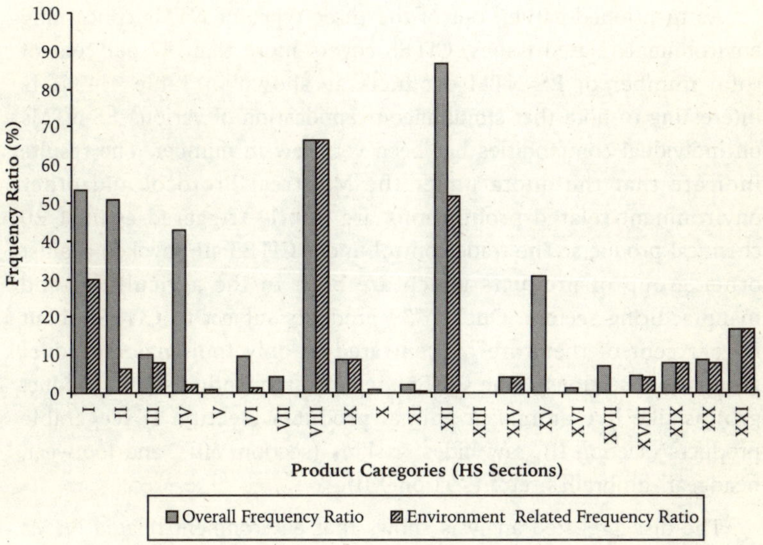

In agriculture, all the broad categories except animal and vegetable fats & oils (HS Section III) are extensively protected by NTBs. In the industrial sector only three broad product categories are strongly protected, and those broad groups are: raw hides & skins, leather, etc. (HS Section VIII); footwear, headgear & umbrella (HS Section XII); and base metals & its associated products (HS Section XV). The recent studies of Mohanty (2001a, 2001b), Mehta and Mohanty (1999) show that most of these products are of to India's export interest to the EU.

Decomposition of overall frequency ratio into environment-related, trade-related and both environmental and trade-related indices shows interesting results. It is observed that against a large number of products, both environmental and trade related NTMs are invoked simultaneously in the agricultural sector. There are about 275 such products which are identified in the agricultural sector. On the contrary, only 197 items in the same sector are subject to only environmental NTMs.

Though environment-related NTMs are WTO-compatible, they are very often being used as an effective instrument to restrict market access to other countries in the EU. Detailed results by HS chapters are presented in Table 4.3.

As mentioned earlier, out of the three types of NTMs concerning environment-related issues, CITES covers more than 97 per cent of total number of ES-NTM products as shown in Table 4.4. It is interesting to note that simultaneous application of various ES-NTMs on individual commodities has been very few in number. The results indicate that the quota under the Montreal Protocol and other environment-related prohibitions are jointly triggered against 20 chemical products. The trade control under CITES are invoked against other group of products which are both in the agricultural and manufacturing sectors. Out of 739 products subject to CITES, about 85 per cent of them are concentrated in only four major product groups (HS sections). The CITES are very frequently used in product groups like live animal & animal products (section I), vegetable products (section II), raw hides & skins (section VIII), and footwear, headgear, umbrellas, etc. (section XII).

The disaggregated analysis shows that environment-related NTMs is mostly concentrated in few broad product groups as identified in Table 4.4. A detailed analysis on the concentration of environment-related NTMs is presented in Table 4.5. As mentioned earlier, agricultural sector is more protected than manufacturing sector, and within the agricultural sector, some broad product segments are more heavily protected than others. Some of the heavily protected areas in the agricultural sector include: live animal (Chapter 1), meat & edible meat offal (Chapter 2), fish & crustaceans (Chapter 3), live trees & other plant bulbs (Chapter 6) & preparations of meat and fish (Chapter 16). Among the non-primary product categories, some of the highly protected product segments by the environmentally related NTMs are raw hides & skins (other than furskins) (Chapter 41), articles of leather (Chapter 42), furskins & artificial fur (Chapter 43), wood & articles of wood (Chapter 44) and footwear, gaiters, etc. (Chapter 64). These products constitute more than 88 per cent of total number of products subject to CITES related NTMs. In total, 11 chapters in the agricultural sector and 18 chapters in the manufacturing sector are affected by environmental NTMs.

The customs tariff of the EU indicates that 87.5 per cent of the total product (national) lines is subject to *ad valorem* tariff. The simple average MFN tariff of the Union is about 5.9 per cent in 1998. The results indicate that specific tariffs are very often triggered for imports in the EU. In every eight-products (at national line), one line is subject to specific tariff. Both in terms of *ad valorem* tariff and specific tariff, agricultural sector is overwhelmingly protected than that of the industrial sector as shown in Table 4.6. The average tariff rate of the agricultural sector was 12 per cent as against 6 per cent for the industrial sector in 1998.[8] It may be noted that all the broad agricultural product groups are strongly protected by unprecedented high tariffs and specific tariffs.[9] Another important observation which emerges from the results is that average tariffs of selected product groups have been very high and several product groups have witnessed high incidence of specific tariffs within these product ranges. However, product groups having high average tariff rate coupled with specific tariffs are mostly seen in the agricultural sector.

The results show that average tariff of broad product groups are not only alarmingly large in agricultural sector but also in the manufacturing sector. Such broad product groups having significant average tariff are: live animals & animal products (section I) vegetable products (section II), animal & vegetable fats & oils (section III), prepared food-stuffs & beverages (section IV), plastics & articles thereof (section VII), textiles (section XI) and footgear, headgear & umbrella (section XII). Similarly, specific tariffs are mostly concentrated in primary products and select manufacturing products, such as chemicals (section VI), articles of stone, plaster and cement (section XIII) and optical, photography and cinematography (section

8. For details about India, see Mohanty (2002).

9. UNCTAD has made an attempt to estimate *ad valorem* tariff equivalence of specific tariffs for the year 1996. The results show that average *ad valorem* tariff equivalence of specific tariff is much higher than average *ad valorem* tariff of similar product groups in the EU. For details see Mehta and Mohanty (1999).

XVIII). A detailed results concerning *ad valorem* tariffs and specific tariffs are presented in Table 4.7. The import of primary sector products is not only protected by tariffs but also by the specific tariffs. Out of 25 chapters in the primary sector, the average *ad valorem* rate of products is more than 10 per cent in case of 12 HS chapters. The products in these chapters constitute about 64 per cent of total number of lines in the primary sector. The manufactured tobacco (Chapter 24) is the product group, which is subject to maximum average tariff of 48.2 per cent in 1998. In the manufacturing sector, average MFN tariffs is very close to 10 per cent in some product groups.

The chapters, where average MFN tariffs exceed 10 per cent are made up fabrics and apparel (Chapter 60-63). Average tariff of these group of products is 11.8 per cent. The product lines in these chapters constitute about 37.3 per cent of the total number of products in textiles and textile articles (Section XI). Some of the non-textile product groups witnessing high tariff (more than 6 per cent) are footwear (Chapter 64), aluminum articles (Chapter 76) and vehicles other than railway or train (Chapter 87). However, some of the chapters have witnessed 'nil' duty for imported products such as ores, slag and ash (Chapter 26), pharmaceutical products (Chapter 30), pulp of wood or of other fibrous material (Chapter 47), vegetable plaiting materials and works of art and collector's pieces (Chapter 97).

In the EU, specific tariffs are combined with environment related NTMs and other NTMs in the agricultural sector as shown in Table 4.8. Except for animal and vegetable fats and oils (Section 3), other broad product groups are subject to both specific tariff and NTMs in the agricultural sector. The environment sensitive NTMs combined with specific tariffs have affected only one sector, i.e. live animals products (Section 1). To be more specific, environmental protection is mostly affecting products like live animal (Chapter 1), meat (Chapter 2) & fish preparations (Chapter 16). For details see Table 4.9.

Conclusions

The EU market is largely protected by both the TBs and NTMs. The agricultural sector is more protected than the manufacturing sector. So far as NTMs are concerned, primary sector is more protected than the manufacturing sector. The environment sensitive NTMs have further intensified the level of protection in the Union. In every four products subject to any type of NTM, one product is subject to environmentally sensitive NTM.

We have used frequency ratio to examine the level of protection. It is found that all broad agricultural products have high frequency ratio except for products like fats and oils.

In the manufacturing sector, some of the important product groups are subject to large number of environment sensitive NTMs and South Asian countries have export interest in those products. Some of these broad product segments are hides & skins, leather, other manufactures, etc. Even the average tariffs for these product segments have been very large as compared to overall average tariff of the Union. The export prospects of South Asia is largely affected by the use of multiple NTMs, peak tariff and specific tariff on specific products on which the region has its current and future export interest.

References

Dean, J.M. (1992). "Trade and the Environment: A Survey of the Literature", in Patrick Low (ed.) *International Trade and the Environment. World Bank Discussion Papers* No.159, World Bank, Washington D.C.

Esty, D. (1994). *Greening the GATT, Trade, Environment and the Future,* Institute for International Economics, Washington, D.C.

Jaffe, Adam *et al.* (1995). "Environmental Regulation and the Competitiveness of US Manufacturing", *Journal of Economic Literature,* 33 (1) March.

Jha, Veena *et al.* (1997). *Trade, Environment and Sustainable Development: A South Asian Perspective,* Macmillan Press Ltd., London and St. Martin's Press Inc., New York.

Manoharan, T.R. Beena Pandey and Zafar Dad Khan (2000). "Trade and Environment Linkages: A Review of Conceptual and Policy Issues", *Discussion Paper RISDCP#3/2000,* Research and Information System for the Non-Aligned and other Developing Countries, New Delhi.

Mehta, R. and S.K. Mohanty (1999). *WTO and Industrial Tariffs: An Empirical Analysis for India,* Research and Information System for the Non-Aligned and Other Developing Countries, New Delhi.

Mohanty, S.K. (2002). *Strategy for Industrial Tariff Negotiations in the Millenium Round: India's Export Interest in Developed Countries and Peak Tariffs.*

Sorsa, Piritta (1994). "Competitiveness and Environmental Standards: Some Exploratory Results", *Policy Research Working Paper, No. 1249,* World Bank.

UNCED (1992). *"Agenda 21, Rio Declaration, Forest Principles",* United Nations.

WTO (1999). *Trade and Environment Bulletin No. 29,* (*WT/CTE/4*), World Trade Organization, Geneva.

Table 4.2

Pattern of Environmental and Trade Related NTMs in the EU: Section-wise Analysis (1996)

Section	Product Description	Total HS Lines	Lines Affected by NTMs	Frequency Ratio (Total)	Frequency Ratio (Env.)	Frequency Ratio (Other)	No. of Lines Covered by Env. NTMs	No. of Lines Covered by other NTMs	
								Trade	Health
I	Live Animals & Animal Products	1159	617	53.2	29.8	41.3	345	69	410
II	Vegetable Products	1293	659	51.0	6.3	48.6	82	628	0
III	Animal or Veg. Fats & Oils	148	15	10.1	8.1	2.0	12	3	0
IV	Prepared Foodstuff & Beverages,	1443	619	42.9	2.3	41.8	33	557	46
V	Mineral Products	251	2	0.8	0.0	0.8	0	2	0
VI	Products of the Chemicals	1796	174	9.7	1.1	8.6	20	154	0
VII	Plastics & Articles Thereof	620	27	4.4	0.0	4.4	0	27	0
VIII	Raw Hides & Skins, Leather	157	104	66.2	66.2	0.0	104	0	0
IX	Wood & Articles of Wood	201	18	9.0	9.0	0.0	18	0	0
X	Pulp of Wood or of Other Fibres	252	2	0.8	0.0	0.8	0	2	0
XI	Textile & Textile Articles	1789	39	2.2	0.2	2.0	4	35	0
XII	Footwear, Headgear, Umbrellas	184	159	86.4	51.6	83.7	95	154	0
XIII	Articles of Stone, Plaster, Cement	271	0	0.0	0.0	0.0	0	0	0
XIV	Natural or Cultured Pearl, Jewellery	69	3	4.3	4.3	0.0	3	0	0
XV	Base Metals & Articles	1255	387	30.8	0.0	30.8	0	387	0

Contd. ...

...Contd. ...

Section	*Product Description*	*Total HS Lines*	*Lines Affected by NTMs*	*Frequency Ratio (Total)*	*Frequency Ratio (Env.)*	*Frequency Ratio (Other)*	*No. of Lines Covered by Env. NTMs*	*No. of Lines Covered by other NTMs*	
								Trade	*Health*
XVI	Mach. & Mechanical Appliances	2350	34	1.4	0.0	1.4	0	34	0
XVII	Vehicles, Aircraft, Vessels	277	20	7.2	0.0	7.2	0	20	0
XVIII	Optical, Photograph, cinem.	446	21	4.7	4.3	0.4	19	2	0
XIX	Arms & Ammunition	25	2	8.0	8.0	0.0	2	0	0
XX	Misc. Manufactured Articles	259	23	8.9	8.1	0.8	21	2	0
XXI	Works of Art Collectors' Pieces	6	1	16.7	16.7	0.0	1	0	0
	Total	**14251**	**2926**	**20.5**	**5.3**	**21.0**	**759**	**2076**	**456**

Source: UNCTAD, Trade Analysis and Information System, Winter 1998.

Table 4.3

Pattern of Environment and Trade Related NTMs in the EU: Chapter-wise Analysis (1996)

HS Chapter	Product Description	Total HS Lines	Lines Affected by NTMs	Frequency Ratio (Total)	Frequency Ratio (Env.)	Frequency Ratio (Other)	No. of Lines Covered by Env. NTMs	No. of Lines Covered by Other NTMs	
								Trade	Health
1	Live Animal	78	64	82.1	66.7	79.5	52	0	62
2	Meat & Edible Meat Offal	336	207	61.6	45.8	50.6	154	6	164
3	Fish & Crustaceans, Molluscs	506	138	27.3	24.9	10.3	126	52	0
4	Diary Produce: Birds, Eggs	216	198	91.7	1.4	90.3	3	11	184
5	Products of Animal Origin	23	10	43.5	43.5	0.0	10	0	0
6	Live Trees & Other Plants Bulb	94	81	86.2	86.2	54.3	81	51	0
7	Edible Vegetables & Certain Roots	366	186	50.8	0.0	50.8	0	186	0
8	Edible Fruits & Nuts: Peel or Melon	476	326	68.5	0.0	68.5	0	326	0
9	Coffee, Tea, Mate & Spices	59	6	10.2	0.0	10.2	0	6	0
10	Cereals	79	52	65.8	0.0	65.8	0	52	0
11	Products of the Milling Industry	89	1	1.1	0.0	1.1	0	1	0
12	Oilseeds & Oleaginous Fruits	82	6	7.3	0.0	7.3	0	6	0
13	Lac, Gums, Resins & Other Vegetable Saps & Extracts	36	0	0.0	0.0	0.0	0	0	0
14	Vegetable Plaiting Materials	12	1	8.3	8.3	0.0	1	0	0
15	Animal or Vegetable Fats & Oils	148	15	10.1	8.1	2.0	12	3	0
16	Preparations of Meat & Fish	192	55	28.6	13.5	24.0	26	0	46

Contd. ...

...*Contd.* ...

17	Sugar & Sugar Confectionery	51	0	0.0	0.0	0.0	0	0	0
18	Cocoa & Cocoa Preparations	42	0	0.0	0.0	0.0	0	0	0
19	Prep. of Cereals, Flour, Starch, etc.	133	0	0.0	0.0	0.0	0	0	0
20	Prep. of Vegetables, Fruit, Nuts, etc.	552	335	60.7	0.0	60.7	0	335	0
21	Miscellaneous Edible Preparations	111	27	24.3	4.5	19.8	5	22	0
22	Beverages, Spirit & Vinegar	248	157	63.3	0.0	63.3	0	157	0
23	Residues & Waste from Food Industries	73	20	27.4	2.7	24.7	2	18	0
24	Tobacco & Manufactured Tobacco	41	25	61.0	0.0	61.0	0	25	0
25	Salt, Sulphur, Earths & Stone Plaster, etc.	99	0	0.0	0.0	0.0	0	0	0
26	Ores, Slag & Ash	40	0	0.0	0.0	0.0	0	0	0
27	Mineral Fuels Mineral Oils & Products	112	2	1.8	0.0	1.8	0	2	0
28	Inorganic Chemicals Compounds, etc.	299	19	6.4	0.0	6.4	0	19	0
29	Organic Chemicals	890	84	9.4	2.2	7.2	20	64	0
30	Pharmaceutical Products	81	0	0.0	0.0	0.0	0	0	0
31	Fertilisers	36	16	44.4	0.0	44.4	0	16	0
32	Tanning or Dyeing Extracts	80	1	1.3	0.0	1.3	0	1	0
33	Essential Oils & Resinoids	59	0	0.0	0.0	0.0	0	0	0
34	Soap, Organic Surface Active Agents	31	0	0.0	0.0	0.0	0	0	0
35	Albuminoidal Substance; Modified	40	0	0.0	0.0	0.0	0	0	0
36	Expolosives: Pyrotechnic Products	10	0	0.0	0.0	0.0	0	0	0
37	Photographic or Cinematographic Goods	65	0	0.0	0.0	0.0	0	0	0
38	Miscellaneous Chemical Products	205	54	26.3	0.0	26.3	0	54	0
39	Plastics and Articles Thereof	517	27	5.2	0.0	5.2	0	27	0
40	Rubber and Articles Thereof	103	0	0.0	0.0	0.0	0	0	0

Contd. ...

...Contd. ...

HS Chapter	*Product Description*	*Total HS Lines*	*Lines Affected by NTMs*	*Frequency Ratio (Total)*	*Frequency Ratio (Env.)*	*Frequency Ratio (Other)*	*No. of Lines Covered by Env. NTMs*	*No. of Lines Covered by Other NTMs*	
								Trade	*Health*
41	Raw Hides & Skins (Other than Furskin)	56	50	89.3	89.3	0.0	50	0	0
42	Articles of Leather, Saddlery & Ham	55	21	38.2	38.2	0.0	21	0	0
43	Furskins & Artificial Fur	46	33	71.7	71.7	0.0	33	0	0
44	Wood & Articles of Wood	180	18	10.0	10.0	0.0	18	0	0
45	Cork & Articles of Cork	9	0	0.0	0.0	0.0	0	0	0
46	Manufactures of Straw, of Esparto, etc.	12	0	0.0	0.0	0.0	0	0	0
47	Pulp of Wood or of other Fibrous Materials	20	0	0.0	0.0	0.0	0	0	0
48	Paper & Paperboard	208	2	1.0	0.0	1.0	0	2	0
49	Printed Books, Newspapers, Pictures	24	0	0.0	0.0	0.0	0	0	0
50	Silk	50	0	0.0	0.0	0.0	0	0	0
51	Wool, Fine or Coarse Animal Hair	72	4	5.6	5.6	0.0	4	0	0
52	Cotton	316	30	9.5	0.0	9.5	0	30	0
53	Other Vegetable Textile Fibres; Paper	58	0	0.0	0.0	0.0	0	0	0
54	Man-made Filaments	119	0	0.0	0.0	0.0	0	0	0
55	Man-made Staple Fibres	186	3	1.6	0.0	1.6	0	3	0
56	Wadding, Felt & Non-wovens; Special	123	0	0.0	0.0	0.0	0	0	0
57	Carpets & Other Textile Floor Covering	56	0	0.0	0.0	0.0	0	0	0
58	Special Woven Fabrics; Tufted Textile	83	0	0.0	0.0	0.0	0	0	0
59	Impregnated, Coated, Textile Fabrics	59	0	0.0	0.0	0.0	0	0	0

Contd. ...

...Contd. ...

60	Knitted or Crocheted Fabrics	49	0	0.0	0.0	0.0	0	0	0
61	Articles of Apparel & Clothing Knitted	193	0	0.0	0.0	0.0	0	0	0
62	Articles of Apparel & Cloth not Knitted	281	0	0.0	0.0	0.0	0	0	0
63	Other Made Up Textile Articles	144	2	1.4	0.0	1.4	0	2	0
64	Footwear, Gaiters & Like; Parts of Article	154	154	100.0	58.4	100.0	90	154	0
65	Headgear & Parts Thereof	14	1	7.1	7.1	0.0	1	0	0
66	Umbrellas, Sun Umbrella, Walking-stick	8	3	37.5	37.5	0.0	3	0	0
67	Prepared Feathers & Down & Articles	8	1	12.5	12.5	0.0	1	0	0
68	Articles of Stone, Plaster, Cement, etc.	65	0	0.0	0.0	0.0	0	0	0
69	Ceramic Products	55	0	0.0	0.0	0.0	0	0	0
70	Glass and Glassware	151	0	0.0	0.0	0.0	0	0	0
71	Natural or Cultured Pearls, Jewellery	69	3	4.3	4.3	0.0	3	0	0
72	Iron and Steel	467	347	74.3	0.0	74.3	0	347	0
73	Articles of Iron or Steel	354	29	8.2	0.0	8.2	0	29	0
74	Copper & Articles Thereof	75	0	0.0	0.0	0.0	0	0	0
75	Nickel & Articles Thereof	17	0	0.0	0.0	0.0	0	0	0
76	Aluminium and Articles Thereof	84	0	0.0	0.0	0.0	0	0	0
77	(Reserved for Possible Future Use)	0	0	0.0	0.0	–	0	0	0
78	Lead & Articles Thereof	13	0	0.0	0.0	0.0	0	0	0
79	Zinc & Articles Thereof	13	0	0.0	0.0	0.0	0	0	0
80	Tin & Articles Thereof	9	0	0.0	0.0	0.0	0	0	0
81	Other Base Materials; Cermets; Articles	84	9	10.7	0.0	10.7	0	9	0
82	Tools, Implements, Cultery, Spoon, etc.	91	0	0.0	0.0	0.0	0	0	0
83	Miscellaneous Articles of Base Metal	48	2	4.2	0.0	4.2	0	2	0

Contd. ...

...Contd. ...

HS Chapter	*Product Description*	*Total HS Lines*	*Lines Affected by NTMs*	*Frequency Ratio (Total)*	*Frequency Ratio (Env.)*	*Frequency Ratio (Other)*	*No. of Lines Covered by Env. NTMs*	*No. of Lines Covered by Other NTMs*	
								Trade	*Health*
84	Nuclear Reactors, Boilers, Machinery	1016	7	0.7	0.0	0.7	0	7	0
85	Electrical Machinery & Equip. & Parts	1334	27	2.0	0.0	2.0	0	27	0
86	Railway or Tramway Locomotives	37	0	0.0	0.0	0.0	0	0	0
87	Vehicles Other than Railway or Tram	162	20	12.3	0.0	12.3	0	20	0
88	Aircraft, Spacecraft & Parts Thereof	38	0	0.0	0.0	0.0	0	0	0
89	Ships, Boats & Floating Structures	40	0	0.0	0.0	0.0	0	0	0
90	Optical, Photographic, Cinematograph, etc.	327	4	1.2	0.6	0.6	2	2	0
91	Clocks and Watches & Parts	88	6	6.8	6.8	0.0	6	0	0
92	Musical Instruments; Parts & Access	31	11	35.5	35.5	0.0	11	0	0
93	Arms & Ammunition; Part & Access	25	2	8.0	8.0	0.0	2	0	0
94	Furniture; Bedding, Mattresses	104	0	0.0	0.0	0.0	0	0	0
95	Toys, Games & Sports, Requisite	86	5	5.8	5.8	0.0	5	0	0
96	Misc. Manufactured Articles	69	18	26.1	23.2	2.9	16	2	0
97	Works of Art, Collectors' Pieces	5	1	20.0	20.0	0.0	1	0	0
	Total	**14250**	**2926**	**20.5**	**5.3**	**17.8**	**759**	**2076**	**456**

Source: UNCTAD, Trade Analysis and Information System, Winter 1998.

Table 4.4

Distribution of Environment-sensitive NTMs in the EU: Section-wise Analysis (1996)

Section	Product Description	CITES	Montreal Protocol (Quota)	Prohibition	CITES and Monetary Protocol	Montreal Protocol and Prohibition	All the Three Types of NTMs
I	Live Animals & Animal Products	345	0	0	0	0	0
II	Vegetable Products	82	0	0	0	0	0
III	Animal or Veg. Fats & Oils	12	0	0	0	0	0
IV	Prepared Foodstuff, Beverages, etc.	33	0	0	0	0	0
V	Mineral Products	0	0	0	0	0	0
VI	Products of the Chemicals	0	0	0	0	20	0
VII	Plastics & Articles Thereof	0	0	0	0	0	0
VIII	Raw Hides & Skins, Leather	104	0	0	0	0	0
IX	Wood & Articles of Wood	18	0	0	0	0	0
X	Pulp of Wood or of Other Fibres	0	0	0	0	0	0
XI	Textile & Textile Articles	4	0	0	0	0	0
XII	Footwear, Headgear, Umbrellas	95	0	0	0	0	0
XIII	Articles of Stone, Plaster, Cement	0	0	0	0	0	0
XIV	Natural or Cultured Pearls, Jewellery	3	0	0	0	0	0
XV	Base Metals & Articles of Base Metal	0	0	0	0	0	0
XVI	Machinery & Mechanical Appliances	0	0	0	0	0	0
XVII	Vehicles, Aircraft, Vessels	0	0	0	0	0	0
XVIII	Optical, Photograph, Cinematographic	19	0	0	0	0	0
XIX	Arms & Ammunition	2	0	0	0	0	0
XX	Misc. Manufactured Articles	21	0	0	0	0	0
XXI	Works of Art Collectors' Pieces	1	0	0	0	0	0
	Total	**739**	**0**	**0**	**0**	**20**	**0**

Source: UNCTAD, Trade Analysis and Information System, Winter 1998.

Table 4.5

Distribution of Environment-sensitive NTMs in the EU: Chapter-wise Analysis (1996)

Section	Product Description	CITES	Montreal Protocol (Quota)	Prohibition	CITES and Monetary Protocol	Montreal Protocol and Prohibition	All the Three Types of NTMs
1	Live Animal	52	0	0	0	0	0
2	Meat & Edible Meat Offal	154	0	0	0	0	0
3	Fish & Crustaceans, Molluscs	126	0	0	0	0	0
4	Diary Produce: Birds, Eggs	3	0	0	0	0	0
5	Products of Animal Orgin	10	0	0	0	0	0
6	Live Trees & Other Plants Bulb	81	0	0	0	0	0
7	Edible Vegetables & Certain Roots	0	0	0	0	0	0
8	Edible Fruits & Nuts: Peel or Melon	0	0	0	0	0	0
9	Coffee, Tea, Mate & Spices	0	0	0	0	0	0
10	Cereals	0	0	0	0	0	0
11	Products of the Milling Industry	0	0	0	0	0	0
12	Oil Seeds & Oleaginous Fruits	0	0	0	0	0	0
13	Lac; Gums, Resins & other Vegetable	0	0	0	0	0	0
14	Vegetable Plaiting Materials	1	0	0	0	0	0
15	Animal or Vegetable Fats & Oils	12	0	0	0	0	0
16	Preparations of Meat and Fish	26	0	0	0	0	0
17	Sugars & Sugar Confectionery	0	0	0	0	0	0
18	Cocoa & Cocoa Preparations	0	0	0	0	0	0
19	Prep. of Cereals, Floor, Starch, etc.	0	0	0	0	0	0
20	Prep. of Vegetables, Fruit, Nuts, etc.	0	0	0	0	0	0

Contd. ...

...*Contd.* ...

21	Miscellaneous Edible Preparations	5	0	0	0	0	0
22	Beverages, Spirit & Vinegar	0	0	0	0	0	0
23	Residues & Waste from Food Industries	2	0	0	0	0	0
24	Tobacco & Manufactured Tobacco	0	0	0	0	0	0
25	Salt, Sulphur, Earths & Stone Plaster, etc.	0	0	0	0	0	0
26	Ores, Slag & Ash	0	0	0	0	0	0
27	Mineral Fuels Mineral Oils & Products	0	0	0	0	0	0
28	Inorganic Chemicals Compounds, etc.	0	0	0	0	0	0
29	Organic Chemicals	0	0	0	0	20	0
30	Pharmaceutical Products	0	0	0	0	0	0
31	Fertilisers	0	0	0	0	0	0
32	Tanning or Dyeing Extracts	0	0	0	0	0	0
33	Essential Oils & Resinoids	0	0	0	0	0	0
34	Soap, Organic Surface Active Agents	0	0	0	0	0	0
35	Albuminoidal Substance; Modified	0	0	0	0	0	0
36	Explosives: Pyrotechnic Products	0	0	0	0	0	0
37	Photographic or Cinematographic Goods	0	0	0	0	0	0
38	Miscellaneous Chemical Products	0	0	0	0	0	0
39	Plastics & Articles Thereof	0	0	0	0	0	0
40	Rubber & Articles Thereof	0	0	0	0	0	0
41	Raw Hides & Skins (Other than Furskin)	50	0	0	0	0	0
42	Articles of Leather, Saddlery & Ham	21	0	0	0	0	0
43	Furskins & Artificial Fur	33	0	0	0	0	0
44	Wood & Articles of Wood	18	0	0	0	0	0
45	Cork & Articles of Cork	0	0	0	0	0	0
46	Manufactures of Straw, of Esparto, etc.	0	0	0	0	0	0
47	Pulp of Wood or of Other Fibrous Materials	0	0	0	0	0	0

Contd. ...

...Contd. ...

Section	Product Description	CITES	Montreal Protocol (Quota)	Prohibition	CITES and Monetary Protocol	Montreal Protocol and Prohibition	All the Three Types of NTMs
48	Paper & Paperboard	0	0	0	0	0	0
49	Printed Books, Newspapers, Pictures	0	0	0	0	0	0
50	Silk	0	0	0	0	0	0
51	Wool, Fine or Coarse Animal Hair	4	0	0	0	0	0
52	Cotton	0	0	0	0	0	0
53	Other Vegetable Textile Fibres; Paper	0	0	0	0	0	0
54	Man-made Filaments	0	0	0	0	0	0
55	Man-made Staple Fibres	0	0	0	0	0	0
56	Wadding, Felt & Non-wovens; Special	0	0	0	0	0	0
57	Carpets & Other Textile Floor Covering	0	0	0	0	0	0
58	Special Woven Fabrics; Tufted Textile	0	0	0	0	0	0
59	Impregnated, Coated, Textile Fabrics	0	0	0	0	0	0
60	Knitted or Crocheted Fabrics	0	0	0	0	0	0
61	Articles of Apparel & Clothing Knitted	0	0	0	0	0	0
62	Articles of Apparel & Cloth Not Knitted	0	0	0	0	0	0
63	Other Made Up Textile Articles	0	0	0	0	0	0
64	Footwear, Gaiters & Like; Parts of Article	90	0	0	0	0	0
65	Headgear & Parts Thereof	1	0	0	0	0	0
66	Umbrellas, Sun Umbrella, Walking-stick	3	0	0	0	0	0
67	Prepared Feathers & Down & Articles	1	0	0	0	0	0
68	Articles of Stone, Plaster, Cement, etc.	0	0	0	0	0	0
69	Ceramic Products	0	0	0	0	0	0
70	Glass and Glassware	0	0	0	0	0	0
71	Natural or Cultured Pearls, Jewellery	3	0	0	0	0	0
72	Iron & Steel	0	0	0	0	0	0

Contd. ...

...*Contd.* ...

73	Articles of Iron or Steel	0	0	0	0	0	0
74	Copper & Articles Thereof	0	0	0	0	0	0
75	Nickel & Articles Thereof	0	0	0	0	0	0
76	Aluminium & Articles Thereof	0	0	0	0	0	0
77	(Reserved for Possible Future Use)	0	0	0	0	0	0
78	Lead & Articles Thereof	0	0	0	0	0	0
79	Zinc & Articles Thereof	0	0	0	0	0	0
80	Tin & Articles Thereof	0	0	0	0	0	0
81	Other Base Materials; Cermets; Articles	0	0	0	0	0	0
82	Tools, Implements, Cultery, Spoon, etc.	0	0	0	0	0	0
83	Miscellaneous Articles of Base Metal	0	0	0	0	0	0
84	Nuclear Reactors, Boilers, Machinery	0	0	0	0	0	0
85	Electrical Machinery & Equip. & Parts	0	0	0	0	0	0
86	Railway or Tramway Locomotives	0	0	0	0	0	0
87	Vehicles Other than Railway or Tram	0	0	0	0	0	0
88	Aircraft, Spacecraft & Parts Thereof	0	0	0	0	0	0
89	Ships, Boats & Floating Structures	0	0	0	0	0	0
90	Optical, Photographic, Cinematgraph, etc.	2	0	0	0	0	0
91	Clocks & Watches & Parts	6	0	0	0	0	0
92	Musical Instruments; Parts & Access	11	0	0	0	0	0
93	Arms & Ammunition; Part & Access	2	0	0	0	0	0
94	Furniture; Bedding, Mattresses	0	0	0	0	0	0
95	Toys, Games & Sports, Requisite	5	0	0	0	0	0
96	Misc. Manufactured Articles	16	0	0	0	0	0
97	Works of Art, Collectors' Pieces	1	0	0	0	0	0
	Total	739	0	0	0	20	0

Source: UNCTAD, Trade Analysis and Information System, Winter 1998.

Table 4.6

HS Section-wise Distribution of Average Tariff in the EU in 1996

Section	Product Description	Average* Tariff (per cent)	Number of Lines	
			Tariff	Specific Tariff
I	Live Animals & Animal Products	9.4	622	537
II	Vegetable Products	9.3	865	428
III	Animal or Veg. Fats & Oils	7.5	138	10
IV	Prepared Foodstuff, Beverages, etc.	17.5	715	728
V	Mineral Products	0.8	247	4
VI	Products of the Chemicals	3.8	1768	28
VII	Plastics & Articles Thereof	6.9	620	0
VIII	Raw Hides & Skins, Leather	3.1	157	0
IX	Wood & Articles of Wood	2.9	201	0
X	Pulp of wood or of Other Fibres	4.5	252	0
XI	Textile & Textile Articles	9.2	1786	3
XII	Footwear, Headgear, Umbrellas	8.7	184	0
XIII	Articles of Stone, Plaster, Cement	4.4	261	10
XIV	Natural or Cultured Pearls, Jewellery	1.0	69	0
XV	Base Metals & Articles of Base Metal	3.6	1255	0
XVI	Machinery & Mechanical Appliances	2.5	2350	0
XVII	Vehicles, Aircraft, Vessels	4.8	277	0
XVIII	Optical, Photograph, Cinematographic	3.0	414	32
XIX	Arms & Ammunition	2.9	25	0
XX	Misc Manufactured Articles	3.5	259	0
XXI	Works of Art Collectors' Pieces	0.0	6	0
	Total	5.9	12471	1780

Note: *Specific tariffs are not included in the calculation. There are about 1780 product lines (national lines) which are subject to specific tariffs.
Source: UNCTAD, Trade Analysis and Information System, Winter 1998.

Table 4.7

HS Chapter-wise Distribution of Average Tariff in the EU in 1996

Chapter	*Product Description*	*Average**	*Number of Tariff Lines*
1	Live Animal	3.1	17
2	Meat & Edible Meat Offal	5.8	66
3	Fish & Crustaceans, Molluscs	10.5	506
4	Diary Produce: Birds, Eggs	6.6	10
5	Products of Animal Orgin	0.4	23
6	Live Trees & Other Plants Bulb	10.4	94
7	Edible Vegetables & Certain Roots	11.1	240
8	Edible Fruits & Nuts: Peel or Melon	11.1	330
9	Coffee, Tea, Mate & Spices	3.3	59
10	Cereals	4.0	8
11	Products of the Milling Industry	13.4	7
12	Oil Seeds & Leoginous Fruits	1.4	79
13	Lac; Gums, Resins & Other Vegetable	9.5	36
14	Vegetable Plaiting Materials	0.0	12
15	Animal or Vegetable Fats & Oils	7.5	138
16	Preparations of Meat & Fish	17.9	175
17	Sugar & Sugar Confectionery	13.3	3
18	Cocoa & Cocoa Preparations	7.6	8
19	Prep. of Cereals, Floor, Starch, etc.	14.8	12
20	Prep. Of Vegetables, Fruit, Nuts, etc.	19.5	405
21	Miscellaneous Edible Preparations	10.4	62
22	Beverages, Spirit & Vinegar	11.8	9
23	Residues & Waste from Food Industries	2.1	33
24	Tobacco & Manufactured Tobacco	48.2	8
25	Salt, Sulphur, Earth & Stone Plaster, etc.	0.3	95
26	Ores, Slag & Ash	0.0	40
27	Mineral Fuels Mineral Oils & Products	1.6	112
28	Inorganic Chemicals Compounds, etc.	4.8	299
29	Organic Chemicals	3.3	885
30	Pharmaceutical Products	0.0	81
31	Fertilisers	3.8	36
32	Tanning or Dyeing Extracts	5.8	80
33	Essential Oils & Resinoids	3.2	57
34	Soap, Organic Surface Active Agents	3.0	31

Contd. ...

... Contd. ...

Chapter	Product Description	Average*	Number of Tariff Lines
35	Albuminoidal Substance; Modified	5.2	30
36	Explosives: Pyrotechnic Products	6.5	10
37	Photographic or Cinematographic Goods	5.0	62
38	Miscellaneous Chemical Products	5.3	197
39	Plastics & Articles Thereof	7.8	517
40	Rubber & Articles Thereof	2.5	103
41	Raw Hides & Skins (Other than Furskin)	2.3	56
42	Articles of Leather, Saddlery & Ham	4.9	55
43	Furskins & Artificial Fur	1.7	46
44	Wood & Articles of Wood	2.8	180
45	Cork & Articles of Cork	3.2	9
46	Manufactures of Straw, of Esparto, etc.	3.3	12
47	Pulp of Wood or of Other Fibrous Materials	0.0	20
48	Paper & Paperboard	5.3	208
49	Printed Books, Newspapers, Pictures	1.6	24
50	Silk	5.5	50
51	Wool, Fine or Coarse Animal Hair	6.9	72
52	Cotton	8.4	316
53	Other Vegetable Textile Fibres; Paper	4.0	58
54	Man-made Filaments	7.9	119
55	Man-made Staple Fibres	8.3	186
56	Wadding, Felt & Non-wovens; Special	6.7	123
57	Carpets & Other Textile Floor Covering	8.1	53
58	Special Woven Fabrics; Tufted Textile	9.0	83
59	Impregnated, Coated, Textile Fabrics	7.2	59
60	Knitted or Crocheted Fabrics	10.2	49
61	Articles of Apparel & Clothing Knitted	12.5	193
62	Articles of Apparel & Cloth Not Knitted	12.4	281
63	Other Made Up Textile Articles	10.6	144
64	Footwear, Gaiters & Like; Parts of Article	9.7	154
65	Headgear & Parts Thereof	3.1	14
66	Umbrellas, Sun Umbrella, Walking-stick	4.6	8
67	Prepared Feathers & Down & Articles	3.4	8
68	Articles of Stone, Plaster, Cement, etc.	1.8	65
69	Ceramic Products	5.3	55

Contd. ...

...Contd. ...

Chapter	*Product Description*	*Average**	*Number of Tariff Lines*
70	Glass & Glassware	5.3	141
71	Natural or Cultured Pearls, Jewellery	1.0	69
72	Iron & Steel	3.0	467
73	Articles of Iron or Steel	4.0	354
74	Copper & Articles Thereof	3.6	75
75	Nickel & Articles Thereof	1.2	17
76	Aluminium & Articles Thereof	6.8	84
77	(Reserved for Possible Future Use)	0	0
78	Lead & Articles Thereof	3.1	13
79	Zinc & Articles Thereof	3.5	13
80	Tin & Articles Thereof	0.5	9
81	Other Base Materials; Cermets; Articles	4.0	84
82	Tools, Implements, Cultery, Spoon, etc.	3.8	91
83	Miscellaneous Articles of Base Metal	2.5	48
84	Nuclear Reactors, Boilers, Machinery	2.0	1016
85	Electrical Machinery & Equip. & Parts	3.0	1334
86	Railway or Tramway Locomotives	2.4	37
87	Vehicles Other than Railway or Tram	6.9	162
88	Aircraft, Spacecraft & Parts Therof	2.0	38
89	Ships, Boats & Floating Structures	1.2	40
90	Optical, Photographic, Cinematograph, etc.	2.7	327
91	Clocks & Watches & Parts	4.2	56
92	Musical Instruments; Parts & Access	3.7	31
93	Arms & Ammunition; Part & Access	2.9	25
94	Furniture; Bedding, Mattresses	2.7	104
95	Toys, Games & Sports, Requisite	4.2	86
96	Misc. Manufactured Articles	3.8	69
97	Works of Art, Collectors' Pieces	0.0	6
	Total	**5.95**	**12471**

Note: * Specific tariffs are not included in the calculation. There are about 1780 product lines (national lines) which are subject to specific tariffs.

Source: UNCTAD, Trade Analysis and Information System, Winter 1998.

Table 4.8

Pattern of Specific Tariffs and Environment-related NTMs in the EU: Section-wise Analysis

Section	Product Description	No. of Lines: Specific Tariff	No. of Lines: Nominal Tariff	Env. NTMs with Specific Tariff	Other NTMs with Specific Tariff	Total No. of Lines
1	Live Animals & Animal Products	537	622	171	381	1159
2	Vegetable Products	428	865	0	307	1293
3	Animal or Veg. Fats & Oils	10	138	0	0	148
4	Prepared Foodstuff, Beverages, etc.	728	715	2	317	1443
5	Mineral Products	4	247	0	0	251
6	Products of the Chemicals	28	1768	0	0	1796
7	Plastics & Articles Thereof	0	620	0	0	620
8	Raw Hides & Skins, Leather	0	157	0	0	157
9	Wood & Articles of Wood	0	201	0	0	201
10	Pulp of Wood or of Other Fibres	0	252	0	0	252
11	Textile & Textile Articles	3	1786	0	0	1789
12	Footwear, Headgear, Umbrellas	0	184	0	0	184
13	Articles of Stone, Plaster, Cement	10	261	0	0	271
14	Natural or Cultured Pearls, Jewellery	0	69	0	0	69
15	Base Metals & Articles of Base Metal	0	1255	0	0	1255
16	Machinery & Mechanical Appliances	0	2350	0	0	2350
17	Vehicles, Aircraft, Vessels	0	277	0	0	277
18	Optical, Photograph, Cinematographic	32	414	0	0	446
19	Arms & Ammunition	0	25	0	0	25
20	Misc Manufactured Articles	0	259	0	0	259
21	Works of Art Collectors' Pieces	0	6	0	0	6
	Total	1780	12471	173	1005	14251

Source: UNCTAD, Trade Analysis and Information System (TRAINS), Winter 1998.

Table 4.9

Pattern of Specific Tariffs and Environment-related NTMs in the EU: Chapter-wise Analysis

Section	Product Description	No. of Lines: Specific Tariff	No. of Lines: Nominal Tariff	Env. NTMs with Specific Tariff	Other NTMs with Specific Tariff	Total No. of Lines
1	Live Animal	61	17	45	48	78
2	Meat & Edible Meat Offal	270	66	126	140	336
3	Fish & Crustaceans, Molluscs	0	506	0	0	506
4	Diary Produce: Birds, Eggs	206	10	0	193	216
5	Products of Animal Origin	0	23	0	0	23
6	Live Trees & Other Plants Bulb	0	94	0	0	94
7	Edible Vegetables & Certain Roots	126	240	0	109	366
8	Edible Fruits & Nuts: Peel or Melon	146	330	0	146	476
9	Coffee, Tea, Mate & Spices	0	59	0	0	59
10	Cereals	71	8	0	48	79
11	Products of the Milling Industry	82	7	0	1	89
12	Oil Seeds & Oleaginous Fruits	3	79	0	3	82
13	Lac; Gums, Resins & Other Vegetable	0	36	0	0	36
14	Vegetable Plaiting Materials	0	12	0	0	12
15	Animal or Vegetable Fats & Oils	10	138	0	0	148
16	Preparations of Meat & Fish	17	175	2	15	192
17	Sugar & Sugar Confectionery	48	3	0	0	51
18	Cocoa & Cocoa Preparations	34	8	0	0	42
19	Prep. of Cereals, Floor, Starch, etc.	121	12	0	0	133
20	Prep. of Vegetables, Fruit, Nuts, etc.	147	405	0	106	552
21	Miscellaneous Edible Preparations	49	62	0	5	111
22	Beverages, Spirits & Vinegar	239	9	0	156	248

Contd. ...

...Contd. ...

Section	*Product Description*	*No. of Lines: Specific Tariff*	*No. of Lines: Nominal Tariff*	*Env. NTMs with Specific Tariff*	*Other NTMs with Specific Tariff*	*Total No. of Lines*
23	Residues & Waste from Food Industries	40	33	0	15	73
24	Tobacco & Manufactured Tobacco	33	8	0	20	41
25	Salt, Sulphur, Earths & Stone Plaster, etc.	4	95	0	0	99
26	Ores, Slag & Ash	0	40	0	0	40
27	Mineral Fuels Mineral Oils & Products	0	112	0	0	112
28	Inorganic Chemicals Compounds, etc.	0	299	0	0	299
29	Organic Chemicals	5	885	0	0	890
30	Pharmaceutical Products	0	81	0	0	81
31	Fertilisers	0	36	0	0	36
32	Tanning or Dyeing Extracts	0	80	0	0	80
33	Essential Oils & Resinoids	2	57	0	0	59
34	Soap, Organic Surface Active Agents	0	31	0	0	31
35	Albuminoidal Substance; Modified	10	30	0	0	40
36	Explosives: Pyrotechnic Products	0	10	0	0	10
37	Photographic or Cinematographic Goods	3	62	0	0	65
38	Miscellaneous Chemical Products	8	197	0	0	205
39	Plastics & Articles Thereof	0	517	0	0	517
40	Rubber & Articles Thereof	0	103	0	0	103
41	Raw Hides & Skins (Other than Furskin)	0	56	0	0	56
42	Articles of Leather, Saddlery & Ham	0	55	0	0	55
43	Furskins & Artificial Fur	0	46	0	0	46
44	Wood & Articles of Wood	0	180	0	0	180
45	Cork & Articles of Cork	0	9	0	0	9
46	Manufactures of Straw, of Esparto, etc.	0	12	0	0	12
47	Pulp of Wood or of Other Fibrous Materials	0	20	0	0	20

Contd...

...*Contd.* ...

48	Paper & Paperboard	0	208	0	0	208
49	Printed Books, Newspapers, Pictures	0	24	0	0	24
50	Silk	0	50	0	0	50
51	Wool, Fine or Coarse Animal Hair	0	72	0	0	72
52	Cotton	0	316	0	0	316
53	Other Vegetable Textile Fibres; Paper	0	58	0	0	58
54	Man-made Filaments	0	119	0	0	119
55	Man-made Staple Fibres	0	186	0	0	186
56	Wadding, Felt & Non-wovens; Special	0	123	0	0	123
57	Carpets & Other Textile Floor Covering	3	53	0	0	56
58	Special Woven Fabrics; Tufted Textile	0	83	0	0	83
59	Impregnated, Coated, Textile Fabrics	0	59	0	0	59
60	Knitted or Crocheted Fabrics	0	49	0	0	49
61	Articles of Apparel & Clothing Knitted	0	193	0	0	193
62	Articles of Apparel & Cloth Not Knitted	0	281	0	0	281
63	Other Made Up Textile Articles	0	144	0	0	144
64	Footwear, Gaiters & Like; Parts of Article	0	154	0	0	154
65	Headgear & Parts Thereof	0	14	0	0	14
66	Umbrellas, Sun Umbrella, Walking-stick	0	8	0	0	8
67	Prepared Feathers & Down & Articles	0	8	0	0	8
68	Articles of Stone, Plaster, Cement, etc.	0	65	0	0	65
69	Ceramic Products	0	55	0	0	55
70	Glass & Glassware	10	141	0	0	151
71	Natural or Cultured Pearls, Jewellery	0	69	0	0	69
72	Iron & Steel	0	467	0	0	467
73	Articles of Iron or Steel	0	354	0	0	354
74	Copper & Articles Thereof	0	75	0	0	75

Contd. ...

...Contd. ...

Section	*Product Description*	*No. of Lines: Specific Tariff*	*No. of Lines: Nominal Tariff*	*Env. NTMs with Specific Tariff*	*Other NTMs with Specific Tariff*	*Total No. of Lines*
75	Nickel & Articles Thereof	0	17	0	0	17
76	Aluminium & Articles Thereof	0	84	0	0	84
77	(Reserved for Possible Future Use)	0	0	0	0	0
78	Lead & Articles Thereof	0	13	0	0	13
79	Zinc & Articles Thereof	0	13	0	0	13
80	Tin & Articles Thereof	0	9	0	0	9
81	Other Base Materials; Cermets; Articles	0	84	0	0	84
82	Tools, Implements, Cutlery, Spoon, etc.	0	91	0	0	91
83	Miscellaneous Articles of Base Metal	0	48	0	0	48
84	Nuclear Reactors, Boilers, Machinery	0	1016	0	0	1016
85	Electrical Machinery & Equip. & Parts	0	1334	0	0	1334
86	Railway or Tramway Locomotives	0	37	0	0	37
87	Vehicles Other than Railway or Tram	0	162	0	0	162
88	Aircraft, Spacecraft & Parts Therof	0	38	0	0	38
89	Ships, Boats & Floating Structures	0	40	0	0	40
90	Optical, Photographic, Cinematograph, etc.	0	327	0	0	327
91	Clocks & Watches & Parts	32	56	0	0	88
92	Musical Instruments; Parts & Access	0	31	0	0	31
93	Arms & Ammunition; Part & Access	0	25	0	0	25
94	Furniture; Bedding, Mattresses	0	104	0	0	104
95	Toys, Games & Sports, Requisite	0	86	0	0	86
96	Misc. Manufactured Articles	0	69	0	0	69
97	Works of Art, Collectors' Pieces	0	6	0	0	6
	Total	**1780**	**12471**	**173**	**1005**	**14251**

Source: UNCTAD, Trade Analysis and Information System (TRAINS), Winter 1998.

5

Case Study of India

SACHIN CHATURVEDI
GUNJAN NAGPAL

Introduction

The debate on the impact of trade liberalisation has intensified with a growing literature on the effects of international trade patterns on the environment. Many participants in the trade and environment debate, in particular, have feared that there could be conflicts between trade liberalisation and environmental safety. The issues of environmental regulation and international competitiveness revolves round the question of harmonisation of standards.[1] It is often feared that competitive deregulation could lead to downward harmonisation of environmental standards. In the context of the WTO, the nature of linkage between trade and environmental measures in promoting sustainable development has become a major concern for the developing countries.

An added dimension to this debate pertains to the national technical regulations and standards pertaining to environment. These environmental standards, often resorted to by the Northern developed countries, are seen as non-tariff barriers against Southern trade. Further, the distinction between environmental standards and health and quality standards is gradually becoming very blurred. Though empirical evidence on this in the literature is extremely limited, some developing countries have experienced losses in exports because of difficulties to comply with certain SPS measures in the import markets.

---

1. Nordstrom *et al.*, (1999), Bhagwati (1996) and Pearson (2000).

Generally, the trade impacts of SPS measures can be grouped into three. First, they can prohibit trade by imposing trade ban on the product or on the inputs used for its production. Second, they can divert trade from one trading partner to another by laying down regulations that discriminate across potential supplies. Third, they can reduce overall trade flows by increasing costs or raising barriers for all potential suppliers.

As the liberalisation of tariff and quantitative restrictions on trade in agricultural and food products has progressed, attention has focussed on technical measures such as food safety regulations, labelling requirements and quality standards. The SPS measures seeks to protect consumers by providing rules for food safety and animal and plant health. In certain cases, stricter SPS measures are applied to imports than domestic supplies. Given the nature and depth of existing regulatory structures in case of SPS in developed countries, developing countries often find it difficult to comply with such standards. At times, it seems that SPS measures may impede trade in agricultural and food products since in many instances they are incompatible with prevailing systems of production and marketing. The developing countries often lack appropriate scientific and technical expertise to deal with such standards. Moreover, the multiplicity of standards in the developed country markets has further compounded the problems faced by developing country exporters.

It was agreed at the WTO Ministerial Conference at Doha that negotiations on issues relating to SPS measures will be addressed on priority basis in the next ministerial conference; however, with failure of Cancun it could not happen. The Committee on Trade and Environment (CTE) was instructed to give particular attention to the effect of environmental measures on market access and trade. This had become relevant in light of the fact that the past decade saw a global proliferation of environment and health related standards along with a rise in the trade in environmentally sensitive goods. Since the inception of WTO, some 2300 notifications have been received and almost 11 per cent of them are related to environment (Nordstrom and Vaughan, 1999). The critics of increasing regulations have protested that stringent environmental regulations force manufactures of pollution-intensive products overseas. It is feared that differences in environmental regulations across the globe is leading to relocation of industries from countries with high environmental process standards to countries with relatively less stringent environmental regulations.

In this chapter an effort is made to examine these emerging issues in the Indian context. Section II takes a stock of trends in the trade of environmentally sensitive goods and India's position in the South Asian context. Section III is an attempt to analyse the whole issue of translocation of dirty industries to India on the basis of available evidence. In the Section IV, we have tried to identify certain emerging challenges while the last section gives the concluding remarks and policy recommendations.

Environment Related Trade Barriers and India

Environmental and health related standards and regulations in the developed country markets have the potential to create barriers to trade. However, now it is clear that the ETBs generally cover all barriers that have been introduced by the importing country to protect the environment, as well as the health and safety of wildlife, plants, animals and humans. The developing countries have had to adjust their production processes in response to the changing environmental regulations in the developed countries. Measures such as pesticide residue levels (MRL) permitted in foodstuffs, emissions standards for machines, and packaging requirements have exerted pressure on the exporters. However, what remains to be seen is the extent of impact of these measures on trade. It is now widely believed that these technical measures impede trade of the developing countries, either implicitly or explicitly.

There are very few studies, which have quantified the impact of ETBs on global trade in general and still fewer for South Asia. The issue of compliance cost is equally concerning for the developing countries. For instance, the costs of upgrading sanitary conditions in the Bangladesh frozen shrimp industry to satisfy EU and US hygiene requirements is estimated to be $ 17.6 million (Cato, 1998). The total industry cost, that is required to maintain Hazard Analysis Critical Control Point (HACCP), is $2.2 million per annum. The European standards are more stringent than HACCP methods. In the case of marine products, EU regulations concerning implementation of food safety systems, additive requirements and other process controls are of very high order. As a result, many of the Indian companies were required to upgrade their facilities, which amounts to a huge expenditure and a number of companies were also forced to close down their factories for a long duration to enable them to upgrade their facilities with heavy investments. Currently, only 90 out of 404 plants in India are approved for fishery exports to EU (Cato, 1998).

Indian Exports

Indian exports have registered a strong growth over the past decade. They have grown from \$ 18,477 million in 1990-91 to \$ 21,588 million in 2001-2002. The growth of exports over the decade was spread across all the major commodity categories. India's exports of agricultural and allied products have increased by about \$40 million from 1998 to 2001 (Table 5.1). However, the share of agriculture in total exports has fallen from about 18 per cent in 1998-99 to 14 per cent in 2000-01. Marine products form a bulk of the exports of agricultural products. More than 3 per cent of India's exports are marine products. In 1999-2001, marine products worth \$ 1183 million were exported. Manufactured goods on the other hand form more than 78 per cent of total exports of India. The export of textile fabrics and manufactures are about 31 per cent of total manufactured goods exports. More than 4 per cent of manufactured goods exports comprises of leather manufactures and almost 11 per cent of it comprise of chemicals and allied products.

Nature and Composition of ETBs

As the liberalisation of tariff and quantitative restrictions on trade in agricultural and food products has progressed, there has been an increased concern about the impact of technical measures especially the environment-related standards on the exports of products. The technical standards such as food safety regulations, labelling requirements and quality and compositional standards have proliferated, particularly in the developed countries.

Table 5.1

Sectoral Profile of Indian Exports

	1998-99		*1999-2000*		*2000-2001*	
	Rs. Cr.	*$ Million*	*Rs. Cr.*	*$ Million*	*Rs. Cr.*	*$ Million*
Agricultural and Allied Products	26104	6205	25016	5773	28535	6246
	18.68	18.68	15.68	15.68	14.02	14.02
Fish & Fish Preparations	4369	1038	5125	1183	6367	1394
	3.13	3.12	3.21	3.21	3.13	3.13
Meat & Meat Preparations	788	187	819	189	1470	322
	0.56	0.56	0.51	0.51	0.72	0.72
Fruits, Vegetables & Pulses	931	221	1247	288	1608	352
	0.67	0.67	0.78	0.78	0.79	0.79
Manufactured Goods	109787	26096	127532	29431	160771	35192
	78.56	78.56	79.93	79.93	78.98	78.98
Textile Fabrics & Manufactures	35581	8457	40178	9272	49831	10908
	25.46	25.46	25.18	25.18	24.48	24.48
Leather & Leather Manufactures	6847	1580	6890	1590	8914	1951
	4.90	4.76	4.32	4.32	4.38	4.38
Chemicals & Allied Products	14211	3378	17389	4013	22850	5002
	10.17	10.17	10.90	10.90	11.22	11.23
Total	**139752**	**33218**	**159561**	**36822**	**203571**	**44560**

Source: Economic Survey (2002), Ministry of Finance, Government of India.

Major issue of concern is that the distinction between environmental, health and quality standards is gradually becoming very blurred. For instance, in the food sector what may be described as quality standard for food may also fall in the category of environmental standards (Box 5.1). Accordingly some of the recent papers have defined ETBs at a very broad level. It covers considerations for protection of environment; protection of wild life; protection of plant health; protection of human health and protection of human safety (Fontagne *et al.*, 2001).

In the last few years, consumer movement across the world, especially in the developed countries, has become very strong. The consciousness for quality products has grown tremendously. This has compelled national governments to take adequate precautions in terms of product specifications for both domestic producers as well as for exporters to these economies.

However, it is very important to distinguish between precautionary and protectionist ETBs. The methodology generally adopted to identify product specific ETBs is on the basis of frequency of its appearance in the list of notifications by member countries (Fontagne *et al.*, 2001). That is, greater the number of countries notifying a particular ETB for a same product, greater the probability of this being a precautionary measure rather than a trade restrictive measure. It is assumed that when a single country or a limited number of countries enforce an ETB it is more likely that these countries have simply enforced a non-tariff barrier. However, when a large number of countries adopt this then the likelihood of this being real threat to

BOX 5.1

Emerging Non-tariff Barriers

Discussions with exporters provided a glimpse of the various hurdles they face. The case in point is the experience of an Indian company, exporting seedless grapes to a large chain of departmental stores in Europe. Before exporting, this company had to fill an elaborate questionnaire, which covered issues like the status of their employees, facilities available to them, and the working conditions. Thus, the exporters are supposed to meet certain social standards before they could start exporting what the importing company is calling as 'socially responsible trading'. This department chain has actually come out with a code for its exporters which apart from social issues, covers building health centres and getting new set of imported instruments for fire extinguishing and evaculation belts, etc. The RIS survey shows that the production cost would go up by 35 to 40 per cent because of compliance with this code.

Apart from this, the grape exporters have to meet the various standards prescribed by 'Eurepgap', which would become mandatory by July 2003. The 'Eurepgap' certification itself requires the exporters to meet a number of conditions, in case of training, planning and preparations, pesticide record keeping, disposal and post harvest preparation. The European countries levy import duty of additional 12.5 pre cent on Indian grapes, whereas a number of African countries do not levy any duty.

Source: Based on *Primary Survey of Industries* conducted by RIS (2001-02).

environment is much greater. The TRAINS database from UNCTAD gives a detailed account of non-tariff trade barriers as all WTO members are supposed to notify non-tariff measures.[2] Annexure A-5.1 gives the list of UNCTAD classification of ETBs. Fontagne *et al.*, (2001) has identified 43 out of 115 effective measures, enforced till 1999 from this database, which may be classified as ETBs. At the global level, selected 185 products have been identified in this paper which face environment related trade barriers at least in one importing country. World imports in these products amounts to US $ 286 billion of which 49 per cent of the total value are affected by ETBs. The distribution by HS classification is very clear, with exception of chemicals and pharmaceuticals, only agro products are the most affected ones (ITC, 2001). In case of India, we have tried to work out India's exports of these 185 commodities. These are largely agricultural commodities and constitute almost 62 per cent of India's total agricultural exports. Out of this 26 per cent goes to United States, 7 per cent to European Union and 5 per cent to Japan.

It is clear from the Annexure A-5.3 that standards are important subjects of contention among WTO members. According to the WTO Secretariat, 193 disputes have been notified to the WTO. Of these, 32 have been settled, 34 Appellate Body and panel reports have been adopted, and 22 cases are active as of 23 May 2000. The rest of the cases are in a consultative phase between affected parties. There have been 25 cases related to TBT or SPS provisions—13 per cent of the whole. In its first year of operation, the DSU saw one fourth—11 out of 44—of its cases referred to these 2 agreements. While the actors have largely been developed countries, the developing countries have also played a role. In fact, the first case resolved in the DSU was brought by a developing country—Venezuela, which won the case—and referenced TBT provisions. Most of these disputes are to do with agriculture and textile sector, and hence are of great interest to the developing countries.

The focus of this section is on selected environmentally sensitive product categories in India, *viz.* agriculture and marine products, leather and textiles related products. The affected sectors due to

---

2. Though many countries do not notify all the NTBs, so this database has to be approached with caution.

external environmental requirements are found to be India's vibrant export oriented sectors such as leather and leather products, textiles, chemicals, marine products, tea and other agricultural products. The nature of NTBs faced by these sectors range from technical standards, product content requirements to mandatory labelling, eco-labelling, packaging requirements and other SPS related measures. Annexure A-5.2 shows India's exports of certain environmentally sensitive products and their share in total exports.

Agriculture and Marine Products

The share of agricultural exports in total exports was 30 per cent in 1990-91, which declined to 22 in 2001-2002. A number of agricultural products of India are facing SPS related problems. Agricultural products are under strict supervision especially in the EU where all imported food products are liable for inspection at the first point of entry for compliance with food laws pertaining to the country of entry. The regulations in the EU also stipulate conditions regarding the labelling of packaging materials used in the imported products. In case of products like peanuts, other nuts and milk, EC has introduced high level of protection by reducing the maximum level of presence of aflatoxin in these products. The level of protection proposed by EU is substantially higher than that provided under Codex recommendations (see details on various standards in Box 5.2). Aflatoxin problem is prevalent in chillies also. Spain recently detained chilli consignments from India. In the case of peanuts, the EU argument has been that the risk involved is of persons contacting cancer in a population of one billion. This is extremely unreasonable because EU population is less than one third of a billion. So the level of SPS protection is not in relation to the extent of risk involved. Though specific measures have been taken in this regard to help out producers and exporters in terms of UNDP supported projects to produce aflatoxin free peanuts (Basu, and Radhakrishanan, 2001). However, wider attempts are yet to be made to internalise the additional cost of compliance in the production structure. The compliance cost for exporters at times is prohibitively high the investment on infrastructure alone ranges from Rs. 12 to 20 lakh for high performance liquid chromatography (Kittu, 2001).

Some of the quarantine restrictions for fresh fruits and vegetables imposed by many countries are also not based on scientific

Table 5.2

Top 10 Indian Agricultural Trade Partners (2001)

Importing Countries	*Value of Imports (US $ Million)*	*Share in Total Indian Agricultural Exports. (%)*
European Union	1865.03	22.23
USA	1051.69	12.54
Japan	770.52	9.18
Bangladesh	468.03	5.58
UAE	458.83	5.47
UK	425.85	5.08
Saudi Arabia	391.9	4.67
Hong Kong	342.08	4.08
Russia	317.74	3.79
Italy	314.93	3.75

Source: *India Trade* Version. 2.0

BOX 5.2

Multiplicity of Standards

One of the major challenges Indian exporters are facing relates to multiplicity of products and other standards. Though international efforts have been made to harmonise standards across the board but lot has yet to be achieved. Recently, Italy and Germany have detained Indian spice consignments on the ground of pesticide residue. Both the above countries failed to convince Indian exporters on the changes they made on their existing regulations on microbial contaminations and contamination due to pesticide residue. This is a blatant denial of facilities offered under Article 7 of the SPS regulations and is causing not only difficulties for India in its regular exports but also leading to loss of opportunities elsewhere.

The efforts of European Spice Association (ESA) to lay down uniform standards and code of practices in collaboration with the spice trade associations of individual European countries are yet to find wider acceptance at EC levels. Until a common European regulation and code of practice is established, traders have to follow regulation of individual countries

Source: Kittu (2001).

justification. Some of the countries are not even acknowledging the statistics in terms of pest and disease prevalence in various parts of the world, submitted by international organisations. This happened in case of India when China imposed a ban on the grapes for a mediterranean fruit fly that does not exist in India.[3]

In Japan, the food sanitation law prohibits the import of many citrus fruits from India without any justification.[4] Indian flower industry is facing a whole set of NTBs while exporting to Japan. In recent past, Japan has imposed zero tolerance clauses on insects, on the assumption that these could possibly be present in Indian flowers. This clause is imposed on particular insects, which are already present in abundance in Japan. There is another problem with regard to quarantine of flowers (APEDA, 2001). The plant quarantine authorities at Japanese airports take a lot of time in the clearance of flower consignments due to elaborate fumigation procedures because of which it takes 5-9 hours to clear a consignment of flowers, which are highly perishable. Many of the South American suppliers of flowers are allowed to do pre-shipment inspection at the port of dispatch. In that case it is possible for Japan to post their inspectors at exit points of flowers. However, the cost of posting inspectors is prohibitively high and would render Indian flowers uncompetitive. Another problem that Indian flower exporters face is that Japanese auction houses bring the Indian roses towards the end of the auction process after entire domestic supply is auctioned and also after flowers from other supplier countries have been auctioned. Since flowers are very perishable, this effects their value in the market. As Table 5.2 shows India's major agricultural trading destines are European Union (22 per cent) and the US (13 per cent) and Japan (9 per cent). In India, there are growing concerns that the benefits of free access to the OECD market may be undermined by non-tariff barriers of stringent and some time arbitrary environment-related regulations.

Food and agriculture form an important part of exports from India. Recently, attempts have been made to widen the range of exports, resulting in the promotion of high-value added items such as processed

3. Personal Communication with S. Dave, APEDA.
4. Based on *Primary Survey of Industries* conducted by RIS.

agro- and marine products on our export basket. There are growing concerns that non-tariff barriers may undermine the benefits of free access to the OECD market stringent and some time arbitrary environment-related regulations. A broader indication of impact of SPS requirements on South Asian exports of agricultural and food products is provided by data on rejections of exports from this region. At present such a data is available only for the United States. Table 5.3 shows that, over the period August 2000 to July 2001, there were significant rejections of imports from South Asia due to microbiological contamination and filth. More than 40 per cent of rejections of exports from India was due to this reason. Food and Drug Administration rejected about 36 per cent of Bangladesh's exports because of microbiological contamination. This shows the considerable problems

Table 5.3

Number of Contravention Cited for US Food and Drug Administration Import Detention, (August 2000-July 2001)

Reasons for Contravention	*India*	*Pakistan*	*Sri Lanka*	*Bangladesh*	*Nepal*
Food Additives	159 (7.4)	12 (1.3)		1 (3.0)	
Pesticide Residues	41 (1.9)				
Heavy Metals	13 (0.6)	4 (0.4)			
Mould	9 (0.4)				
Microbiological Contamination	329 (15.3)	49 (5.5)		12 (36.4)	
Decomposition	7 (0.3)				
Filth	568 (26.4)	12 (1.3)	2 (11.1)	12 (36.4)	
Low Acid Canned Foods	87 (4.1)	25 (2.8)	9 (50.0)	3 (9.1)	
Labelling	338 (15.7)	50 (5.6)	3 (16.7)	1 (3.0)	
Others	597 (27.8)	744 (83.0)	4 (22.2)	4 (12.1)	1
Total	**2148**	**896**	**18**	**33**	**1**

Note: Figures in parentheses gives the percentage share.

Source: US Food and Drug Administration Import Detention Report, 2001.

that South Asian countries have in meeting basic food hygiene requirements. The Table also shows that these countries also have a problem meeting the stringent labelling requirements of the United States. More than 15 per cent of total agricultural imports from India and Sri Lanka were rejected because of their failure to meet these requirements. Other than that inadequate food additives, presence of pesticide residual and heavy metals and low acid canned foods are commonly cited reasons for contravention. Out of 18 import detentions of Sri Lanka, 9 detentions are because of low acid content in the canned food. More sophisticated monitoring and testing facilities, and therefore more costly procedures, are required for meeting these regulations. On top of that, the cost of rejection at the border can be considerable, as it includes loss of product value, transport and other export costs and product re-export or destruction.

Table 5.4

Relative Importance of Environmentally Dirty Products in World Trade

(Value of Exports Expressed in US$ Billions)

Year	*Total Trade*	*Metalliferous Ores*	*Manufactured Metal Products*	*Chemicals*	*Pulp & Paper*	*Iron & Steel*	*Non-ferrous Metals*
1980	2001.9	31.8	37.4	141.3	8.79	75.6	49.9
1985	2103.5	23.8	41.6	177.8	9.61	74.0	36.5
1990	3427.9	36.1	66.1	300.4	14.39	108.0	67.5
1995	5228.4	45.5	104.8	487.5	17.9	145.7	106.1
1998	5400.3	45.6	110.7	512.3	16.6	145.1	108.0

Source: United Nations, *International Trade Statistic Yearbook*, various issues.

Marine Products

Marine products are considered to be the most environmentally sensitive products in the international market. In India, till late seventies, the export of marine products mainly consisted of dried items like dried fish, dried shrimp, shark fins and fish maws, etc. However, later there was a decline in the export of dried marine products, and subsequently the exports of processed items continued to make steady progress in the marine trade. The markets for Indian marine foods were mainly confined to Singapore, Sri Lanka, Myanmar, etc. Afterwards, when the frozen and canned items increasingly figured in the exports basket, USA, France, Canada, Japan and Australia became the important markets for Indian marine products. During 1980s canned items slowly disappeared and frozen items became the prominent ones in India's seafood trade. USA was the principal buyer of Indian frozen shrimp followed by Japan and Western European countries. The demand from Japan included headless shell or shrimp, USA demanded peeled shrimp meat while the European countries preferred the IQF (individually quick frozen) shrimp and the cooked form. During 1998-99, Japan continued to be the single largest buyer of Indian marine products accounting for 22.21 per cent in volume and 49.61 per cent in value, whereas USA accounted for 11.38 per cent of volume and 13.34 per cent of the value of marine products exported. Due to import liberalisation, during the same period, South East Asian countries also emerged as another important market account for 56.57 per cent in volume and 24.25 per cent in value. As against this, the share of member countries in the European countries was 17.91 per cent by volume and 14.80 per cent by value. Therefore, the export of marine products had grown significantly as one of the important item of India's export from a few million US$ in 1961-62 to US$ 1106.9 million in 1997-98 accounting for approximately four per cent of the total export from India (United Nations, 2001).

Among the ETBs faced by seafood, shrimps form the major imports from India pertaining to the level of pesticides and antibiotics. Various antibiotics and chemicals like oxolinic acid and oxytetracyclines without any given limit are totally banned. Consignments containing DDT, Aldrin and Heptachlor are bound to be rejected. The EU directive has also imposed process standards requiring hygiene during handling, processing and storage of marine products. The ban by the US on Indian shrimp products was a unilateral restriction on environmental reasons. In 1996, US banned shrimps from entry unless harvested by aquaculture caught with turtle excluding devices, or by manual instead of mechanical means or in cold water. US lost the case at WTO when India and other affected countries challenged the ban. However, the ban since 1996 adversely affected Indian shrimp exports.

Textiles

In India, textile exports account for 55 per cent of countries export earnings. The share of readymade garments in the export basket has hovered around 12 per cent. In 1999-2000 it was 12.9 per cent then was 12.5 per cent in 2000-2001. This went up marginally to 13 per cent in 2000-2001 but then came down to 11 per cent in 2001-2002. It is also the country's net foreign exchange earner as the import content in textile is very little as compared to the other major export products. This sector shares 4 per cent of GDP and 14 per cent of industrial products. It provides employment to 15 million people. Although the Indian textile industry is characterised as power loom, handloom and mill sectors, in India, majority of the readymade garment exports come from SMEs, which accounts for as much as 63 per cent to the total exports from the textile sector.

Around 40 per cent of India's textiles directed to the EU but nowadays the stringent environment conditions in the developed countries pose adverse consequences in India's export performance. In 1993, Germany textile industry introduced two eco-labels, *viz.* *Markenzeichen Schastoffgeprufth Textilien* (MST), which gave norms for consumer goods and indicates a lower content of pollutants, and *Markenzeichen Unweltschonende Textilien* (MUT), a label which sets norms for production processes. It indicates that all processing conditions are analysed with reference to the degree of pollution of air, water and soil. Apart from this, other national and private labels are also developed in Europe. For example, the *Ostereichisches Textil-Forschungsinstitut* has developed the *Oekotex* label which relates to both raw material and final product. The European Largest Textile and Apparel Companies (ELTAC) has also undertaken an eco-label initiative. In the textile and clothing sector, Germany has recently introduced the *Eco-Tex* Standard–100 which lists various criteria for evaluating textiles from an ecological perspectives. Eco-tex standard-100 has been developed by International Association for Research in Eco-Textiles that offers manufacturers for garments and textiles the opportunity for certification of eco-friendly products.

The use of certain dyestuffs such as cobalt blue and sulphur black has been totally banned in the international market. Though viable substitutes have been explored, but switching over to them again costs higher investment of over US$ 13 million, mainly for the upgradation of the technology and new treatment plants in order to obtain the requisite quality. Likewise opting for non-benzidine dyes, *viz.* direct black 38 dye and direct black 22 was priced at between $8-10 per kg. Though SMEs contribution to the export basket of India are not available, one study does estimate that over 60 per cent of production is by small scale enterprises. One can therefore extrapolate that an important part of exports comes from small-scale units.

Leather Industry

India being the largest holding of livestock in the world is expected to play a dominant role in the leather industry. Leather industry is spread over organised as well as unorganised sector. The small-scale enterprises (SMEs) account for over 75 per cent of the total production. This industry has been identified as one of the thrust areas of exports. Traditionally, the Indian leather industry has been an exporter of tanned hides and skins. Currently leather footwear, footwear uppers, leather goods, garments and other leather goods are among the fastest growing export items in India. Footwear sector has been identified as an area of extreme focus. Export from leather sector today account for about 4.3 per cent of India's exports. The export performance of the leather sector has improved considerably during the past decade. The value of exports went up from Rs. 3,036 crore during 1991-92 to Rs. 6,436 crore in 1998-99. Germany is the largest single export market for Indian leather exports. The other important markets are France, United Kingdom and Italy. The composition of the Indian leather industry is so vast as it employs about 1.4 million people. There are about 2000 tanneries, in different Indian states, *viz.* Tamil Nadu, West Bengal, Maharashtra, Uttar Pradesh, Karnataka and Rajasthan. Out of the total tanneries, 75 per cent of them fall within the SMEs.

On environmental standards, leather industry faces problems from both domestic and external front. The restrictions on the use of certain chemical dyestuffs and several other mandatory regulations in major export markets pose serious problems to the leather sector. Germany along with the other European countries have already restricted the benzidine containing dyes and openly demands benzidine free leather imports. Germany has banned all the imported leather products containing Pentachlorophenol (PCP) with the aim to protect its consumer from the possible incidence of cancer. PCP has been in use as a preservatives for raw hides and skins. It also restricts the use of formaldehyde. Only products treated with environment friendly chemicals are accepted in Germany and other developed country markets. EC standards for PCP in this regard is high around 1000 ppm or 10mg/kg. The stringent environmental standards imposed by EC has been criticised not only by the developing countries but also certain EU member countries themselves such as Italy and France. It is also to be noted that Italy has a big leather industry where as France is itself a major producer of PCP. Therefore, it is reasonable to believe that the purpose of protest from these two countries on PCP ban is to protect their domestic industries.

Most of the studies on competitiveness and environmental standards are based on environmental capital costs. Parikh *et al.* (1994) analysed the effects of domestic environmental policies of foreign countries on India's exports focussing on two sectors: leather and shellfish. The results indicated that the increasingly stringent export standards have contributed to a rise in the cost of production, especially in the leather sector, where costs using the more environmentally friendly methods are nearly three times higher.

However, India, in its efforts to increase its leather exports, banned the production of PCP and accordingly its use in the processing of leather. Government of India has already enforced several environmental regulations such as the Water Act of 1974, the Air Act of 1981 and the Environment Act of 1986 to address environmental problems in the country. For compulsory provisions for effluent treatment, government has set up common effluent treatment (CET) in new tanneries. Large number of alternatives of PCP exist such as 'Busan 30', which is acceptable to the international market. Though in India, PCP was manufactured locally at a cheaper price, whereas 'Busan 30' need to be imported either from Germany or US. According to some recent estimates, the price of this substitute is around 10 times higher than the existing price of PCP. For the easy availability of the substitutes, The Government of India reduced the import duty from 150 to 50 per cent on these chemicals. Further, Centre Leather Research Institute (CLRI) has also indigenously developed technology to enable local production of TCMTB which is a substitute to PCP.

Relocation of Dirty Industries

Environmental regulations have proceeded at different pace in different countries of the world. These differences are particularly pronounced between industrialised countries and developing countries. Widespread concerns have been expressed recently about the relationship between international competitiveness of environmentally sensitive goods and environment regulations. Stringent regulations in the developed countries are hypothesised to lead to industrial flight from the developed nations to developing nations.

The developing countries are concerned about how differences in national environmental regulations might induce investors in the developed countries to shift the location of dirty industries to developing countries, which have relatively lower standards. The justification behind this view is that strong environmental regulations increase the production costs. As a result it is in a firm's interest to locate its production facilities in a country with lower production costs, that is, with relatively lax regulations. This argument focusses solely on the cost effect of environmental regulations on polluting industries, and presumes that the production cost differentials are sufficient inducement for a firm to relocate its production site. If there is a strong incentive to relocate pollution intensive production from countries with strict regulations to countries with lax regulations, this will result in a shift of composition of production in developing countries towards more polluting industries. The flight of polluting industries may also cause economic problems such as unemployment in the short run for the country exporting capital and may also expedite environmental degradation of host countries.

There are a number of studies available in the literature on the impact of environment on industrial relocation. Majority of the studies addressed the issue, in the context of developed countries, particularly the United States. These studies have generally defined the polluting industries on the basis of pollution abatement cost and other expenditures. The definition of dirty industries is crucial to the analysis, but varies from one study to another. This problem is not that serious, as several different definitions seem to yield similar lists of dirty industries.

There is a persistent argument in the literature on trade and environment that differential environmental standards result in the relocation of 'dirty' industries in developing countries and expedite environmental degradation in these countries. It is believed transnational locations blatantly by-pass the more stringent environmental regulations in North by setting up their production plants in the third world countries. But the fact is that this is not always the case. In fact, there are many other factors, such as infrastructure, supply of natural resources and availability of cheap labour that influence the choice of a certain location or investment. Multinationals tend to move to locations where cheap labour, abundant raw materials and good transport system and other facilities are available. Environmental costs, which usually forms a relatively small part of total production costs, seldom effect the location decision of the plant.

The Environmental Regulation and Trade

It is important to analyse the share of environmentally sensitive goods in international trade and trends in that share over time. For our analysis we take five most commonly used environmentally dirty products, namely, manufactured metal products, industrial chemicals, iron and steel, pulp and paper, and non-ferrous metals. Table 5.4 examines the relative importance of dirty goods in global trade. The Table shows, trade in environmentally sensitive product, for selected years.

In 1980, the environmentally sensitive goods accounted for about 17 per cent of total world trade. The relative importance of these goods declined marginally in the period 1997-98. The ferrous and non-ferrous metal accounted for about 36 per cent of trade in these goods and also is a source of over three-quarters of total decline in the world trade share. The share of manufactured metal products has risen marginally from 10 per cent in 1980s to 11 per cent in 1998. The share of paper and pulp manufactures in total trade remained virtually static in the period 1980-1998. Table 5.5 attempts to determine if there are important changes in the geographic origins of these goods in the period 1980-90. It shows the origin of dirty industry from developed and developing countries. It also provides a further breakdown for developing countries into South America,

Eastern Europe and South Asia and the breakdown for developed countries into US and EU.

In case of all environmentally sensitive commodities that we have selected, the share of exports in developed countries has fallen but the share of total exports by developing countries has risen. The developed nations share in value of exports of metalliferous ores accounted for $19.5 billion in 1980, whereas the developing countries accounted for $10.6 billion. By 1998 the developing countries' share increased sharply by 12 per cent and the developed countries' share reduced by eight points. The manufactured metal products are largely exported from the developed countries. They account for approximately 90 per cent of total exports in 1980 and by 1998 the share of exports

Table 5.5

Growth Rates of Exports in Selected Environmentally Sensitive Products

Year	*Developed Economies*	*EEC*	*USA*	*Developing Economies*	*South America*	*Eastern Europe*	*South Asia*
	Metalliferous Ores						
1980-82	-26.2			-20.8	-8.9	257.6	
1982-84	7.6			-8.3	-15.2	-18.6	33.3
1984-86	-9.0			1.3	5.1	-82.3	25.0
1986-88	30.5	80.0	55.6	39.7	29.3	158.8	60.0
1988-90	17.4	14.3	16.7	12.8	24.5	150.0	-9.4
1990-92	-12.0	-9.7	-30.6	-4.1	-3.0	-87.3	6.9
1992-94	10.5	10.8	8.8	7.6	6.2	-19.0	9.7
1994-96	31.9	25.0	13.5	33.1	29.4	138.2	47.1
1996-98	-12.6	-7.8	-16.7		8.0	-19.8	-12.0
Annual Average Growth	2.1	..	..	3.4	4.2	26.5	8.9
	Manufactured Metal Products						
1980-82	6.8			923.1			
1982-84	-4.4			33.6			
1984-86	0.6			-87.7			
1986-88	21.4	24.8	46.2	51.5	45.7	26.2	59.1
1988-90	30.5	33.7	41.0	17.9	38.7	-24.5	17.1
1990-92	10.8	10.4	20.4	26.9	70.0	-2.5	25.6
1992-94	2.8	-4.8	24.6	28.0	29.4	-7.7	32.0
1994-96	25.8	43.7	30.9	30.2	36.4	-16.7	21.3
1996-98	4.3	1.5	14.2	10.5	30.0	30.0	6.1
Average Growth	5.5	9.1	9.8	57.4	20.8	0.4	13.4

Contd. ...

...Contd. ...

Year	*Developed Economies*	*EEC*	*USA*	*Developing Economies*	*South America*	*Eastern Europe*	*South Asia*
			Chemicals				
1980-82	-9.2			3.4			
1982-84	10.4			21.7			
1984-86	21.7			50.0			
1986-88	41.6	40.3	41.6	56.0	42.9	0.0	85.3
1988-90	18.8	18.4	24.3	35.9	34.0	27.3	33.6
1990-92	11.1	9.0	12.9	23.9	16.4	0.0	41.7
1992-94	38.8	11.5	17.3	34.9	26.9	-7.1	41.1
1994-96	3.6	32.2	20.0	32.9	28.3	30.8	34.7
1996-98	1.9	4.2	10.0	5.6	11.8	5.9	4.4
Average Growth	7.7	9.6	7.0	14.7	13.4	4.7	20.1
			Paper and Pulp				
1980-82	-10.1			14.0	36.0	-42.9	-12.7
1982-84	30.6			4.1	14.7	175.0	23.2
1984-86	4.9			3.9	17.9	36.4	-3.1
1986-88	12.9	-2.6	16.3	6.6	26.1	6.7	4.9
1988-90	2.1	-4.6	-8.0	5.3	10.3	18.8	-5.3
1990-92	-1.0	0.0	-2.2	4.2	4.7	-36.8	5.9
1992-94	4.1	9.5	8.9	14.5	23.9	225.0	17.6
1994-96	10.9	11.3	13.9	13.0	1.2	66.7	24.4
1996-98	-2.7	-8.0	-0.7	-0.2	9.5	3.1	-5.5
Average Growth	2.9	0.5	1.6	3.6	6.3	23.6	3.5
			Iron & Steel				
1980-82	-13.6			27.9			
1982-84	-7.0			36.4			
1984-86	10.1			14.7			
1986-88	29.6	30.7	110.0	76.7	67.1	102.6	94.6
1988-90	10.9	18.5	57.1	16.4	20.5	9.1	9.7
1990-92	-3.1	-7.7	12.1	11.3	0.0	-36.9	29.1
1992-94	6.6	2.5	0.0	22.8	9.5	-34.0	31.4
1994-96	12.8	33.7	59.5	38.0	29.0	54.3	48.5
1996-98	-0.6	-1.5	-3.4	0.3	-5.6	27.8	2.0
Average Growth	2.5	6.3	13.1	13.6	10.0	10.2	17.9
			Non-ferrous Metals				
1980-82	-37.8			-29.9			
1982-84	17.2			8.0			
1984-86	0.0			-3.7			
1986-88	60.7	47.8	125.0	91.0	93.9	80.0	123.8
1988-90	11.4	20.6	41.7	12.1	26.6	13.6	-10.6
1990-92	-5.7	-8.1	-3.9	1.2	-8.6	-35.9	33.3
1992-94	7.0	4.9	4.1	23.1	6.8	-33.9	51.8
1994-96	24.4	36.7	33.3	64.4	26.6	20.5	48.2
1996-98	0.0	-1.9	0.0	0.3	-10.0	31.9	1.6
Average Growth	4.3	8.3	11.1	9.2	11.3	6.4	20.7

Source: TRAINS, 2001, UNCTAD, Geneva.

have fallen to 73 per cent. In case of chemicals, iron and steel, and non-ferrous metals, there has been sharp fall in share of developed nations over 1980-98. Meanwhile, the share of developing countries has increased greatly in these environmentally sensitive goods. The paper and pulp goods are the only dirty products that have shown only marginal rise in the share of exports of developing countries.

Over the year 1980-1998, the annual average growth rate of all selected environmentally products is higher in the developing countries than in the developed countries. In case of manufactured metal products, for instance, the annual average growth rate is 5.5 per cent in the developed countries but the developing countries have an annual average growth rate of 57.4 per cent. Among the developing countries this high annual growth rate in export of manufactured metal products is attributed to South America and South Asia. The export of chemicals, which is one of the most pollution intensive product, has also increased in the developing countries over time. The annual average growth of the developing countries in the export of chemicals is about double the annual growth rate of exports by the developed countries. Most of this growth is attributed to South Asia, which has an annual growth rate of 20.1 per cent. Iron and steel and non-ferrous products are also considered to be highly pollution intensive as well. In the period 1980-1998, the annual growth rate of exports in iron and steel products in developing countries was 13.6 per cent, whereas the annual growth rate of developed countries was 2.5 per cent. Similarly, the annual growth rate of exports in non-ferrous products was 4.3 per cent in the developed countries and 9.2 in the developing countries.

The trends provide a clear evidence of relative decline in importance of environmentally sensitive products in industrial countries' exports, while increases in the relative importance of these products in case of developing countries, especially South America and South Asia. It should be borne in mind that these increases in the relative importance of dirty industry trade have taken place against a marginal reduction in the share of environmentally dirty goods in total trade.

The Environment Regulation and Investment

The governance of international capital flows will be one of the key environmental policy issues of the next decade. Given the lack of effective state regulation in developing countries, the explosion of foreign direct investment (FDI) has triggered a hot debate about how to govern global investment. The debate revolves around the environmental impacts of FDI and whether the globalisation would induce environmental standards to harmonise upwards or downwards. In recent years, some economists have been worried that developing countries may seek to attract FDI by competitively undercutting each other's environmental standards, i.e. race to bottom. The 'dirty industry' migration hypothesis suggests that the countries with less stringent environmental standards attract the increasing fraction of FDI into polluting industries (Leonard, 1988). It is important to examine the extent of FDI in case of developing nations to conclude that the migration of dirty industries into developing countries is taking place.

The 1990s witnessed a sea change in the pattern of international capital flows. In 1990, official sources accounted for more than half of international capital flows to developing countries. The growth of FDI has been equally dramatic. The volume of FDI flows to developing countries has nearly quadrupled from 1990-95. The Table 5.6 gives the total FDI flow into the South Asian region for the period year 1994-99. The share of FDI to India was 0.3 per cent of the total FDI to the world in 1994. However, the share of FDI to India has gone down to 0.25 per cent in 1999. While the volume of FDI in the world has grown at the average rate of 27 per cent, the volume of FDI into India has also increased at the same rate over the period 1994-99. The volume of FDI into Bangladesh has grown at a much faster rate over the same time period.

To study if the foreign investments have moved into polluting industries it is important to examine the sector-wise break-up of FDI into India. Since it is difficult to get hold of actual figures, we use the approvals instead (Table 5.7).

On examining the sector-wise distribution of FDI it is observed that over the last decade there is some shift in the pattern of foreign direct investment flows. In 1992, electrical equipment, fuels, chemicals and food-processing were the industries that received the bulk of FDIs. However, telecommunication, drug and pharmaceutical industries have received a large share of total FDI's into India. Fuels and electrical equipment continue to attract high level of foreign direct investment but, chemicals and food processing industries are no more receiving a

Table 5.6

FDI Inflows into the South Asian Region (1994-99)

(in Millions of Dollars)

	1994	*1995*	*1996*	*1997*	*1998*	*1999*
World	255988	331844	377516	473052	680082	865487
Bangladesh	6	13	11	141	308	150
India	973	2144	2426	3577	2635	2168
Pakistan	419	719	918	713	507	531
Nepal	7	8	19	23	12	132
Sri Lanka	166	66	133	435	206	202

Source: United Nations, *World Investment Report, 2000.*

high level of FDI. The total FDI approvals into the telecommunication sector has risen from 1324 million to 94631 million rupees. Similarly, there has been a sharp increase in the share of drugs and pharmaceuticals industries in the total FDI. The volume of FDI into food processing industries declined from 4400 in 1992 to 2755 million rupees. The chemical industry, which is considered to be one of the most polluting industry, also experienced a decline in the investment flows from Rs. 5020 million in 1992 to Rs. 2727 million in the year 2000. From the data it is difficult to imply that the FDI has increased in the pollution intensive industry. As some of clean industries, such as telecommunication and drugs and pharmaceuticals have attracted large volumes of FDI in the recent years.

The 'dirty industry' migration hypothesis suggests that the countries with less stringent environmental standards attract the increasing fraction of FDI into polluting industries (Leonard, 1988). To test this hypothesis for India we correlated the sector-wise distribution of FDI[5] into India in the period 1991-2000 with pollution intensity of that sector. Pollution intensity is defined as the amount in weight of pollutants released by an industry per unit production of that industry. Since data on sector-wise pollution content is not available in India, the pollution intensity data from United States Environmental Protection Agency's toxic release[6] database is used. It

5. Source of FDI data is Department of Industrial Policy and Promotion, Ministry of Commerce and Industry, Government of India.
6. The TRI is a database, which provides information to the public about releases of toxic chemicals from various industries into environment.

Table 5.7

Percentage Share of Sector-wise FDI Approvals in Total FDIs (1992-2000)

	1992	*1993*	*1994*	*1995*	*1996*	*1997*	*1998*	*1999*	*2000*
Metallurgical Industries	1.73	14.66	18.08	1.18	6.52	4.58	7.20	4.94	4.81
Fuels	31.51	8.02	17.61	24.68	16.69	48.15	44.54	20.00	15.18
Electrical Equipments	10.65	4.14	8.55	3.41	8.51	4.00	4.63	8.98	33.54
Telecommunications	3.08	0.12	0.97	55.18	11.93	12.99	10.06	13.75	24.89
Transportation Industry	3.85	7.43	8.20	3.95	-3.03	14.10	5.07	21.94	2.68
Industrial Machinery	2.69	6.49	6.62	0.24	0.53	0.40	0.35	0.95	0.12
Fertilisers	0.02	0.01	0.04	0.01	0.67	0.00	0.00	0.00	0.00
Chemicals (Other than Fertilisers)	11.67	9.36	11.67	3.85	7.78	5.15	6.43	2.86	0.72
Photographic Raw Film & Paper	0.18	0.03	0.07	0.03	0.00	0.36	0.00	0.03	0.02
Dye Stuffs	0.00	0.04	0.04	0.06	0.01	0.01	0.24	0.02	0.00
Drugs & Pharmaceuticals	0.72	0.60	0.76	0.58	0.33	0.33	0.30	0.28	4.28
Textiles (Included, Dyed Printed)	3.57	4.08	6.11	0.97	1.14	1.08	0.75	1.14	0.61
Paper & Pulp Including Paper Product	0.47	0.81	0.82	0.98	2.90	1.25	0.35	2.42	0.43
Fermentation Industries	0.12	0.84	0.76	1.93	0.23	0.46	0.00	0.01	0.04
Food Processing Industries	10.23	4.09	9.77	1.04	9.49	3.63	2.05	0.50	0.72
Vegetable Oils & *Vanaspati*	0.17	0.09	0.15	0.04	0.01	0.26	0.00	0.17	0.00
Soaps, Cosmetics & Toilet Prep.	0.44	0.24	0.32	0.05	0.07	0.30	0.20	0.00	0.00
Rubber Goods	0.08	0.19	0.39	0.11	0.97	0.09	1.78	0.44	0.03
Leather, Leather Goods & Pickers	0.65	0.08	0.48	0.12	0.08	0.11	0.16	0.09	0.06
Glass	0.03	0.89	1.25	0.38	0.85	0.85	0.63	1.50	0.05

Source: *United Nations World Investment Report*, 2000.

is acknowledged that industrial pollution intensity is location sensitive and varies from country to country (Ferrantino and Lindins, 1999). However, since we are using it to examine the correlation between FDI flows and dirtiness of industries in case of India, this does not pose any major setback to the study.

The Figure 5.1 gives the comparison of amount of FDI into various sectors in India and their pollution intensity.

If the 'industrial flight' hypothesis was valid, then large FDI would correspond to industries with high pollution intensity. However, it can be seen from the Figure 5.1 that while some highly polluting industries received relatively very low amount of FDI, some relatively cleaner industries such as food processing, electrical equipment and telecommunication received high level of FDI. The statistical correlation between FDI flows into India and the pollution content is

Figure 5.1

Correlation between Pollution Intensity and Foreign Direct Investment (1991-2000)

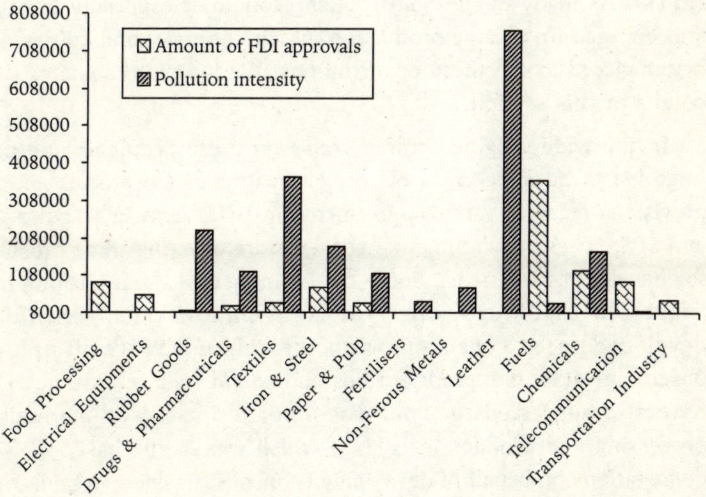

–0.36. The negative correlation indicates that FDI is occurring in relatively clean industries over this decade.

The FDI data provides little empirical evidence to suggest that differences in the stringency of environmental process standards may induce industries located in high standard jurisdiction to relocate to low standard jurisdictions in order to preserve their competitive position

Emerging Policy Challenges

In last decade or so environmental concerns have proliferated, encompassing a large section of trade. This emanates from different international commitments such as Cartagena Biosafety Protocol and Kyoto Protocol. Apart from these Protocols, there are some emerging issues which are posing severe policy challenges to the governments in the developing countries, for instance the growing electronic trade and the subsequent waste being generated out of it. There is also a debate about the GMOs and their position in the international trade. However, along with these challenges are the emerging opportunities in terms of growing market size of environment friendly products (EFPs). As many of the South Asian economies especially India are still engaged in 'organic production' of agricultural commodities, it is better placed to tap these opportunities. We look into some of these details in this section.

In the wide ranging commentaries on these protocols one finds huge list of possible areas of conflicts with these protocols and the provisions made in various agreements in WTO especially under TBT and SPS.[7] These environmental concerns are affecting trade prospects for developing countries. Some of the industrial organisations have suggested that the Kyoto Protocol's empowerment should be developed to be consistent with the existing WTO disciplines, observing MFN, like product rules and sound science criteria.[8] They have also suggested to avoid discrimination based on methods of processing and production (PPMs) which was at the basis of Kyoto deliberations on behalf of developing countries. Business organisations are also concerned about some forms of domestic implementation

7. For details see Zarrilli, 2000, Nielson *et al.*, 2000.

8. Please see the Policy Statement from International Chamber of Commerce, October 26, 1999.

which seeks to offset 'Carbon leakage' could lead to the establishment of trade barriers in the form of tariffs, border taxes, discrimination and labelling. There are several views about trade of goods, which would be affected by Kyoto Protocol. For instance, the study by Montgomery *et al.*, 2000 suggest that compliance with the Kyoto Protocol would result in loss of economic welfare to the tune of $ 900 million and the cost will not be limited to developed countries. The adoption of Clean Development Mechanism (CDM), which allows developed countries to invest in low cost energy reductions in developing countries, which may reduce cost of compliance. It also refers to an emerging division among developing countries themselves as countries like China and India export energy intensive goods and benefit from energy price increases in developed countries.

Growing e-Trade[9]

The electronic industry was traditionally thought to be a relatively clean industry. But this image has taken a dent with greater realisation of ecological hazards emanating from the sector. The ecological hazards come not only from the processes of electronic commodity production but there are also some increasing worries about the disposal of electronic waste. In this regard, the EU Directives are expected to come into force in 2008 and this will have serious implications on the production processes of component manufactures, exporting to the European Union. The electronic industry is one of the fastest growing industries today. It has emerged as one of the major sectors in India's export basket in recent years. The electronic industry has achieved a cumulative annual growth of 20 per cent in production and over 40 per cent in exports till 1998. The exports from this industry have been growing slowly over the past few years. Total electronic hardware exports were Rs. 11,080 million in the year 1993-94 and went upto Rs. 18,000 million in the year 1998-1999 and then took a dip to Rs. 14,000 million in the year 1999-2000. The Indian electronic sector comprises a few large companies, some SMEs and a large number of tiny and household enterprises.

Electronic waste is unwanted computers, monitors, televisions, audio equipment, printers, and other home electronic devices. Technological advances for all electronic equipments, particularly

9. Data discussed under e-trade is taken from: SVTC, 2001; IE, 2002; OECD, 1999.

computers, quickly render them obsolete. More than 20 million personal computers became obsolete in 1998 in EU. Only about 14 per cent were reused or recycled, while most were disposed of or remain in storage. Electronic equipment contains metals and other materials that can become hazardous to human health and the environment if they are not properly managed. Each year EU ends up with 6 million tonnes of electronic waste. The waste stream of electronic and electric equipments has been identified as one of the fastest growing waste streams in the EU constituting 4 per cent of municipal waste today, increasing by 16-28 per cent every 5 years. There are increasing concerns in EU about the electronic waste disposal. In order to reduce the amount of electrical and electronic wastes disposed, the European Commission has adopted a proposal on Waste Electrical and Electronic Equipment (WEEE) and a proposal for Directives on the restriction of the use of certain hazardous substances in electrical and electronic equipments. The proposed Directives are designed to tackle the fast increasing waste stream of electronic and electrical equipments. The Directives would limit the total waste going to final disposal. Producers will be responsible for taking back and recycling electronic and electrical equipments. In order to prevent the generation of hazardous waste, the proposal for Directives requires the restriction of use of certain hazardous substances. The wide-ranging directive would require equipments manufacturers to accept and recycle equipments including computers, mobile phones, televisions, refrigerators and washing machine, under 'polluter pays' principle. The use of substances such as heavy metals, lead, mercury, cadmium, and hexavalent chromium will be banned and these required to be substituted by various types of brominated flame retardants. All these substitution requirements are to be met by the producers by 1 January 2008.

The implications of these emerging environmental regulations for companies in the developing countries that export to European markets are potentially very deep. The companies may have to redesign their manufacturing processes. A recent study shows that so far there is very little awareness amongst the Indian electronic component manufactures about the EU directives and its implications. A lack of environmentally sound technologies could be disastrous for this sector. It would become impossible to sustain production and, hence, trade if ecologically friendly practices are not imbibed. A lack of information about the various requirements and standards in importing countries could pose very serious hurdle to trade.

Labelling and Genetically Modified Goods

In recent past biotechnology has posed some serious policy dilemmas before developing counters. European Union, Japan, Australia and several other countries have launched the plans for mandatory labelling of genetically modified (GM) foods. EU members have recently voted to require labels for food containing more than one per cent of GM material. The Canada and the US have complained to the WTO regarding European measures to label GM products as labelling may violate WTO rules. Concern has been expressed by the US that vigilance is needed to ensure the EU plan is not used to bar US imports. In Australia, the deferment of a government decision regarding the labelling of GMOs is indicative of the uncertainty regarding conflict between eco-labelling and WTO jurisdiction (Chaturvedi, 2004).

It is worth recalling here that the Cartagena Protocol on Biosafety (CPB) was negotiated under the auspicious of Convention on Biological Diversity (CBD) in 1992. The Protocol was adopted by a large number of countries (64) in 2000. This Protocol provides rules for safe transfer, handling, use of, and disposal of living modified organisms (LMOs). The wide objective of the Protocol is to address the threats posed by LMOs to biological diversity along with to human health. There are three major areas of concerns which are generally found to be conflicting with the spirit and provisions in the SPS/TBT agreement under WTO.

In terms of its spirit, the SPS Agreement seems to be restrictive in nature while the Biosafety Protocol empowers for even taking grand measures for protection. Though the SPS agreement covers a wide spectrum of issues concerning human health which may affect access for trade of GMOs while the Biosafety Protocol apart from being GMO specific talks of biodiversity and health in general. The whole understanding of precautionary principle under Article 5.7 of SPS and Article 11.8 of Biosafety Protocol is contradictory in nature. Another area of concern is the acceptable level of risk which may be allowed while trading GMOs. On the areas of risk assessment and management SPS broadly sets the tone for acceptable level of risk at the international level being endorsed by any international institution while Biosafety Protocol refers to an exclusivist approach which may be adopted at national level.

Though in last decade the trade in biotechnology products has grown many fold but due to lack of adequate classification of such products there is hardly any evidence from the South Asian region to substantiate this perception. The current estimated biotechnology market size for India is $1475 million (*Economist*, 2001). The product range from biotechnology related instruments, drugs and even agricultural and food products containing transgenic traces, commercial field trials of which, has not been permitted in many South Asian countries.[10]

However, in case of US a conscious effort has been made to develop a concept of collecting trade data on Advance Technology Products (ATP) by the US Census Bureau. All of the biotechnology products on the ATP list appear to belong to biologics. This is largely of therapeutic products derived from living organisms which include vaccines, human blood, plasma, proteins and monoclonal antibodies. This definition does not match with the definitions evolved by other countries. The OECD in a separate exercise is attempting to evolve biotechnology statistics at least at the level of OECD countries (OECD, 2001). The growing resistance within Europe of GM crops has already affected US agricultural exports. Soyabean, where 35 per cent of US output is composed of genetically modified varieties exports to the EU, declined from 9 million tonnes in 1997 to 6 million tonnes in 1998. Similarly, exports of corn (maize) in 1997 from US to EU was 1.6 million tonnes which has now declined to 0.3 million tonnes (Perdikis, *et al.*, 2004). The global market for transgenic crops and related products has grown very fast in the last quinquennium. The sales have increased from $ 75 million in 1995 to $ 3 billion in 2000. This is likely to reach $ 25 billion by 2010 (Zarrilli, 2000).

It is fairly possible that a conflict between SPS Agreement and Biosafety Protocol may come up in not such a distant future. Though

10. In fact Sri Lanka had to withdraw an official order banning imports of all food items containing GMOs. See *RIS Biotechnology and Development Review* 4(1) for more details.

they have emerged in two different settings but addresses similar issues in contradictory terms. SPS is to address health issues in a wider context while biosafety protocol addresses health in a narrow context of trade in GMOs. The SPS imposes a restrictive regime emanating from international product standards setting institutions while Biosafety Protocol allows member countries to evolve their own necessary measures to protect their health and environment. Therefore, the Protocol has not laid any provisions for addressing disputes while, SPS, being mandatory in spirit, has a backing from a strong Dispute Settlement Provision (DSP).

Eco-labelling Measures

Along with statutory eco-standards, criteria are also being set in importing countries, through eco-labelling. In EU, several labelling systems already exist which apply to the final output as well as the process and production method itself. An eco-labelling scheme informs consumers that a labelled product is environmentally superior than other products in the same category. It is a form of third-party certification, with the certification done by a purely private organisation such as Green Seal in the United States, or by a mixed public entity, as in the Canadian System. Eco-labelling schemes help consumers make decisions about the products they buy and whether they are environmentally friendly. There are several existing eco-label schemes around the world including the German Green Spot, the Nordic Swan and the US Green Seal.

Most current eco-label programmes are 'cradle to grave', that is, they involve some form of analysis based on the environmental consequences of their manufacture, use and disposal. Eco-labelling schemes already in place, may conflict with the WTO Agreement on TBT. WTO rules decree that countries should not discriminate between products based on the environmental consequences of production techniques (Jha *et al.,* 1999). The eco-labelling issue blends into two other market access obstacles, *viz.* environmentally motivated government procurement and ISO 14000 series.

Discriminatory government procurement is a longstanding non-tariff trade barrier (NTB) regulated to some extent by the WTOs Government Procurement code. In recent years some governments have added environmental criteria to their procurement policies. The question arises is whether such criteria constitute a new set of NTBs. While green-consumerism in public procurement is not inherently against the interests of exporting countries, the potential for covert trade discrimination and the likelihood of increased costs for exporters are real, especially for smaller suppliers in developing countries. The 14000 series of ISO can also potentially act as an NTB. In 1993, the ISO extended its work to environmental management systems (EMS) in its 14000 series. The ISO 14000 initiative mirrors the broader trade issues of technical regulations, technical standards, and eco-labelling schemes. Though its purpose might be improvement of environmental performance and facilitation of international trade, but unless carefully crafted, the 14000 series can act as a trade barrier. Specifically, if the EMS is prescriptive, detailed and costly to comply with, and if compliance requires strict criteria for sub-contractor and suppliers' performance, it will be difficult for firms in developing countries to qualify. If ISO registration becomes a condition for at least some types of international business, the trade barrier potential may be realised.

In order to address the potential trade concerns, activities are taking place in various fora such as WTO, OECD, ISO and the Global Eco-labelling Network (GEN). The WTO is determining in what ways does eco-labelling fall within the scope of the WTO's Agreement on TBT. GEN is pursuing cooperation as well as the possible harmonisation and mutual recognition among programmes. Mutual recognition could involve trading partners accepting each other's established eco-labelling criteria, consequently lowering or eliminating the probability that the criteria would act as barriers to trade among nations. Other types of mutual recognition might include reciprocal acceptance of credibility or the acceptance by one programme of another's test procedures and facilities.

As Table 5.8 explains Germany was the first country to establish an eco-labelling progamme in 1977 and now there are about 25 eco-labelling programmes around the world. Along with the proliferation of such programmes a concern has arisen regarding their potential impact on international trade. Although the motivating force behind eco-labelling programme is often domestic environmental improvement, manufactures are concerned that their access to foreign markets may be restricted by the environmental criteria set by programmes in those markets.

Just as technical standards may create non-tariff trade barriers, eco-labelling schemes may inadvertently or deliberately create impediments to trade. Product categories eligible for labels may be selected to favour domestic over foreign producer's interest. The criteria for granting labels may also favour domestic over foreign producers. One example of this is the German textile-labelling scheme, The Blue Angel, which virtually prefers products, which use artificial dyes produced in Germany to natural dyes. Discrimination may exist in product classification system. Austria imposed a mandatory labelling requirement for tropical timber and products but did not impose similar requirements for temperate forest products. This was later withdrawn after objections from the ASEAN countries. Another way that these eco-labelling requirements may act as a non-tariff trade barrier is the cost of compliance. As with technical regulations, the cost of compliance may be much higher for small export suppliers. They may have to incur cost such as information costs, plant inspection and certification cost and acquisition of approved inputs. In general, the criteria for granting eco-labels under life cycle assessment may be based on environmental conditions in the importing country and may be inappropriate in the country of production and export.

Indian Eco-Mark

In India, the Government launched the eco-labelling scheme known as 'Eco-Mark' in 1991. "Any product which is made, used or disposed off in a way that significantly reduces the harm it would otherwise cause the environment could be considered as environment-friendly product." The criteria of Indian eco-mark are based on cradle-to grave approach, i.e. from raw material extraction to manufacturing and to disposal. There are three stages involved in the eco-mark scheme. The first stage is to determine product categories for the scheme, which is done by a steering committee, set up in the Ministry of Environment and Forests. The committee also formulates strategies for promotion, implementation, future development and improvements in the working of the scheme. The second stage involves the identification of a specific product and recommendation of individual criteria to be adopted. A technical committee set up in the Central Pollution Control Board (CPCB) carries this out. Finally,

Table 5.8

Details of Various Eco-labelling Programmes

Eco-labelling Programmes	*Country*	*Year of Establishment*	*Products*	*Cost of Acquiring*
Blue Angel	Germany	1977	Paints & Varnishes, Paper, Cleaning Products, Construction Material, Textiles	DM 300
Nordic Swan	Sweden, Norway, Finland, Iceland	1989	Paper Products, Detergents, Cleaning Agents	US$ 2000 + additional testing and verification costs.
Environmental Choice	Sweden	1990	Paper Products, Detergents, Cleaning Agents	5000 SEK
EU Eco-label Award Scheme	EU	1990	Detergents, Cleaners, Paper and Textiles	US$ 660
Environmental Choice Programme	Canada	1988	Paper Products, Construction & Home Materials	$CAN 10,000
Green Seal	USA	1990	Paper & Newsprint, Engine Oil Products, Paints, Reusable Bags	—
Eco-Mark	Japan	1996		
NF Environment	France	1991		20,000 FF

Source: OECD, *Eco-labelling: Actual Effects of Selected Programmes*, 1997.

the Bureau of Indian Standards (BIS) assesses and certifies the products and draw up a contract with the manufactures.

The Indian eco-mark has not achieved the desired results so far. The response from Indian companies for the award of Eco-mark was very poor. In 1998, M/s Madhya Bharat Paper Mills Limited successfully completed the formalities and received Indian Eco-mark license for producing writing and printing paper. While eco-labelling in other countries are gaining popularity, the initiatives taken by the Government of India still waiting for a breakthrough. The Government has initiated several steps to tackle the above problems. For instance, a scheme encouraging industries to produce environment–friendly products is in the pipeline. Similarly, in the case of leather products, the pre-condition for ISI mark for obtaining eco-mark is not required for those who are producing a certificate from the buyers in the foreign market. With these initiatives, the number of companies approaching for eco-mark is likely to be increased. The problem of obtaining multiple eco-labelling will be solved through agreement of mutual recognition of eco-labelling schemes across countries.

However, during the primary data collection exercise at RIS, there were frequent references to the option of the eco-labelling programme of the importing country awarding its own eco-label to products meeting the process-related criteria of the exporting country and the use and disposal criteria of the importing country. In this regard, the life-cycle analysis for eco-labelling could be split into 'cradle to export border' and 'import border to grave'. While mutual recognition could benefit countries, which already have national eco-labelling programmes, in other cases the concept of 'equivalencies' was seen as more useful. Establishing such equivalence would take account of comparable environmental objectives, different ways of achieving them and differences in environmental and developmental conditions across countries.

Opportunities for Environment Friendly Products (EFPs)

It is difficult to define Environment Friendly Products (EFPs) as no product is absolutely environment-friendly. Therefore, it is necessary to ensure that environmental claims are credible to take account of consumer interests and support fair competition. Eco-labelling and promotion of EFPs to take advantage of public concerns over environment could generate market opportunities, but could also create barriers to trade. Eco-criteria based on environmental and technological conditions in the importing country might imply significant costs for foreign producers and might also be environmentally less appropriate in the context of the local conditions in the foreign country of production.

The constraints in the export of environmentally friendly products from India has been discussed in Saqib and Kaushik (2001). Though organic farming has a tremendous scope in India, there exist a number of bottlenecks for farmers who are interested in organic farming and for the industry who want to process and export these products. These impediments are in the area of production, marketing and infrastructure. Some of the small growers find certification as a major problem. Standards are too high and are creating unfair barrier to production and trade. India has no local certification systems for organic products and farmers have to depend on foreign certification like IFOAM and SKAL. This is very expensive and is feasible for big holdings only.

The United States is the largest single-country market for organic foods, with \$4.2 billion in sales for 1997. As Table 5.9 shows the 1997 organic food market in the EU is estimated to be worth \$4.5 billion (Segger). In Europe, Germany (\$1.6 billion), France (\$508 million), and the United Kingdom (\$445 million) have the largest organic retail sales. Consumer commitment to organics is strong throughout the EU, with 20 per cent to 38 per cent regularly or occasionally purchasing organic foods. Retail price premiums in Europe average from 10 per cent to 50 per cent above conventional products. Import shares are highest in Germany and the United Kingdom, which are major food processors, and in the Netherlands, which is a primary re-exporter. Retail sales are lower in Canada (\$68 million) and Australia (\$60 million), although both countries are active in exporting organics—Australia to Asia and Canada to the US and Europe. Price premiums in Canada average 30 per cent, but a range from 12 per cent to 65 per cent is found across the states of Australia. Import share is very high in Canada, purchased mostly from the US, and very low in Australia, consistent with the organic share of total food sales in each country. Japan (\$1.7 billion) and

China ($1.2 billion) offer large retail markets, with negligible quantity currently supplied by imports. Price premiums in these countries are similar to those in the EU, averaging 15 per cent to 30 per cent. Market participation rates are also similar, between 27 per cent and 36 per cent.

Table 5.9

Market Size for Organic Products

Country	*Approx. Retail Value (US $)*	*Year*[a]	*Organic Share of Total Food Sales (%)*	*Import Share of Organic Sales (%)*	*Average Retail Price Premium (%)*
Austria	270 million	1997	2.5	30	20-30
Belgium	75 million	1997	1.0	50	20
Denmark	190 million	1997	<3.0	25	15-40
France	508 million	1996	0.4	10	25-35
Germany	1.6 billion	1997	1.5	60	30
Netherlands	230 million	1997	1.5	60	15-20
Sweden	200 million	1997	2.0	30	15-50
United Kingdom	445 million	1997	2.0	70	0-30
Canada	68 million	1995	1.0	80	30
Australia	60 million	1995	0.2	0-13[b]	12-65[b]
China[c]	1.2 billion	1995	6.0[d]	0	30
Japan[c]	1.7 billion	1997	1.0	1	20-30

Note: a – Year given is for retail value data; b – Varies by state; c – In this country, organic includes 'low chemical'; d – Based on production value, not retail sales.

Source: *American Journal for Agriculture Economics*, 80(5), 1998: 1125-1129.

Policy Recommendations and Concluding Remarks

In recent past, several issues related to environment have emerged which have serious bearings on multilateral trading agreements. It is important to clear the relationship between eco-labelling and TBT agreement. In this regard, exploration of the possibility of government participation in such schemes may be a relevant exercise. This may help in ensuring greater transparency in the working. It is important that for transparency automatic notification to all parties should be sent and, in fact, for trade purposes, the *ex-ante* transparency is especially important. The implications of the TBT agreement extended beyond transparency question may include important substantive elements such as equivalency, mutual recognition, dispute settlement and technical assistance. However, though India has launched an eco-labelling programme, certainly increased assistance in this area is needed, with focus on technical assistance to help develop and improve certification procedures.

India would also have to upgrade the national system for testing, certification and laboratory accreditation so as to be at par with the global trade demands. In this regard, it is important to focus on capacity building in the private sector. Another related aspect is to closely study the emerging pattern of Mutual Recognition Agreements (MRAs) already underway between couple of OECD countries. The MRAs would start the process of removing duplicative testing and certification requirements in a number of product sectors.

As is clear for various studies the exports are facing a number of institutional constraints in meeting the international standards. Apart from this, lack of timely and precise information can be an obstacle to trade. In this regard, small and medium enterprises face innumerable problems. Whilst large firms obtain timely and accurate information directly from the importers in developed country markets and various other sources, SMEs tend to depend on secondary sources, basically government sources, often implying considerable time delays. Creating awareness of regulations and voluntary labels and available eco-friendly technology would require government intervention.

The certification of environmentally friendly products may be costly and confusing, in particular when producers in developing countries depend on the testing and certification bodies in the developed countries. The creation of standardisation bodies or the expansion of existing bodies in developing countries and steps contributing to their international recognition are of key importance. Improvements in environmental infrastructure play an important role in reducing the costs of compliance. In India, many government, trade, regulatory and research entities have some responsibility for addressing such measures, but there is no one entity directing and coordinating overall government effort. Some entities' role and responsibilities regarding these measures are not clearly defined, and these entities have had difficulty coordinating their activities. As far as testing facilities are concerned, our laboratories are poorly equipped in machines and in skilled manpower, which has led to poor reputation of our test results in international markets. Thus exporters resort to multinational testing facilities which are more expensive. Accreditation of laboratories is voluntary and without any accountability. This has led to a mushrooming up of laboratories that are inappropriately equipped and often resort to issuing fake certificates.

There is limited support from the Government of India available to the exporter to meet the increasing requirements. For instance, Agricultural and Processed Food Products Export Development Authority (APEDA) provides some assistance to the food exporters. They are given assistance for purchase of specialised transport machinery, setting up of pre-cooling facilities with proper air handling system, provision of facilities for pre-shipment treatment such as fumigation, X-ray screening, hot water dip treatment, setting up of vapour heat treatment, electronic beam processing, assistance for setting up of environment control system and setting up of specialised storage facilities such as high humidity cold storage. These limited assistance schemes are subject to a number of restraints and some of these schemes are also being put to an end. Therefore, a more dynamic and vibrant strategy is required to face ETBs and help out the industry on this front.

As is clear, producers of organic products face several sets of potential constraints related to conversion, production, marketing and government support policies. Constraints on conversion to certified organic agriculture in developing countries include uncertainty about markets and price premiums. Certification costs, technical requirements and SPS measures might act as obstacles to exports of organic food products from developing countries. Furthermore, organic vegetable and fruit markets tend to rely largely on locally produced food. In this regard governments should support the development of the agriculture sector, in particular the production of high-value products, improvement of quality and better participation in global value chains, by taking fiscal and other measures to attract investment and technologies into new areas. Options may also be explored for providing WTO acceptable support to the agriculture sector, in particular for research and development and quality assurance especially for reducing costs of certification of organic producers in developing countries by setting up local certification systems, promoting small holder certification, and reducing the costs of international accreditation for certifies in developing countries. In this context there is extreme urgency to develop international mechanisms to develop information channels to provide market information and analysis about these products and strengthen capacity-building initiatives.

The emerging trade and environment debate is highly complicated and inter-disciplinary in nature. In India this becomes much more complicated as the number of agencies dealing with different commodities is very large. Moreover, the agencies supposed to be doing the coordination often loose track of affected trade as at micro level the magnitude is too small to be taken into account. And there is no mechanism in place to pool data at one place so as to facilitate putting pieces take together and develop a macro-vision out of it. Therefore, there is an immediate need to develop a database in India, profiling trade and environment related cases. Once a sizable set of data is available, attempts may be made to quantify the trade distortion effects of those environment measures, if at all they are there.

References

APEDA (2001). Personal Communication with Mr. S. Dave, Director, Agricultural and Processed Food Products Export Development Authority (APEDA), New Delhi.

Basu, M.S. and Radhakrishanan (2001). "Capacity Building in Peanut Farms: A Success Story", paper presented at the *UNCTAD/World Bank Workshop*.

Bhagwati, J. and R. Hudec (1996). *Fair Trade and Harmonisation-Prerequisites for Free Trade?*, The MIT Press.

Bharucha (1994). "Impact of Environmental Standards and Regulations on India's Exports", study by UNCTAD.

————. (2000). "The Impact of Environmental Standards and Regulations Set in Foreign Markets on India's Exports", in Veena Jha, G. Hewison and M. Undenhills (eds.) *Trade, Environment and Sustainable Development: A South Asia Perspective*, Macmillan Press, London, pp. 123-42.

Cato, J.C. (1998). "Economic Issues Associated with Seafood Safety and Implementation of Seafood", *HACCP Programmes*, FAO, Rome.

Chaturvedi, S. (2004). "WTO, Biosafety Regulatory Regime and Trade in Genetically Modified Goods: Options before Developing Countries—An Indian Perspective", in Sachin Chaturvedi and S.R. Rao (eds.), *Biotechnology and Development: Challenges and Opportunities for Asia*, Academic Foundation, New Delhi.

Ferrantino, M.J. and Linda A. Lindins (1999). "The Effect of Global Trade Liberalisation on Toxic Emissions in Industry", *Weltwirtschaftliches Archiv Reivew of World Economics*, Band 135 (1), pp. 128-55.

Fontagne, Lionel, Friedrich von Kirchbach and Mondher Mimouni (2001). "A First Assessment of Environment-Related Trade Barriers", *mimeo*, October.

Han, Ki-Ju and John B. Braden (1996). *Environment and Trade: New Evidence from U.S. Manufacturing*, University of Illinois, Department of Economics, *mimeo.*

IE (2002). "Not just Scrap, Delhi Gets its Share of E-Waste Too", *Indian Express*, March, 1.

ITC (2001). "Environmental Trade Barriers: Who Wins Who Losses, Whats the Score?" *International Trade Forum*, August 3.

Jha, Veena, Anil Markandya, Rene Vossenaar (1999). *Reconciling Trade and the Environment: Lessons from Case Studies in Developing Countries*, Edward Elgar, UK.

Kalt, J.P. (1998). "The Impact of Domestic Environmental Regulatory Policies on US International Competitiveness", in A.M. Spence and H.A. Hazard, (eds.), *International Competitiveness*, Ballinger, Cambridge, M.A, Harper and Row, pp. 221-62.

Kittu, C. (2001). "Issues on SPS and Environmental Standards for India", paper presented at the *UNCTAD/World Bank Conference* on 11-13 January.

Leonard, H.J. (1988). *Pollution and the Struggle for the World Product: Multinational Corporations, Environment, and International Comparative Advantage*, Cambridge University Press.

Letchumanan, R. (1998). "Trade, Environment and Competitiveness: Testing the 'Pollution Haven' Hypothesis from Technology Perceptive", *UNU Working Paper* No. 43.

Montgomery, David and Paul Bernstein (2000). *Kyoto's Impact on Trade and Growth*, Cooler Heads Coalition, December 16.

Nielsen, *et al.* (2000). "GMOs, Trade Policy and Welfare in Rich and Poor Countries", *Policy Discussion Paper No. 0021*, Centre for International Economic Studies, University of Adelaide, Australia.

Nordstrom, H. and S. Vaughan (1999). "Trade and Environment", *SPECIAL Studies 4*, The World Trade Organization, Geneva.

OECD (1999). "Emerging Market Economy Forum", Conference on *Foreign Direct Investment and the Environment*, OECD, The Hague, Netherland, January 28-29.

————. (2001). "Biotechnology Statistics in OECD Member Countries: Compendium of Existing National Statistics", *STI Working Paper No. 6*.

Parikh, J.K., V.K. Sharma, U. Ghosh and M.K. Panda (1994). *Trade and Environment Linkages: A Case Study of India*, report prepared for United Nations Conference on Trade and Development, August.

Pearson, Charles (ed.) (2000). *Multinational Corporations, Environment, and the Third World*, Duke University Press, Durham, NC.

Saqib and Kaushik (2001). "Exporting Environment Friendly Products: Scope for Organic Farming in India", *RGICS Working Paper Series No. 21*.

SVTC (2001). *Europe Proposes Producer 'Take-Back' of Electronic Waste*, Silicon Valley Toxics Coalition.

United Nations (2001). *International Trade Statistics Yearbook*.

Zarrilli, S. (2000). "WTO Sanitary and Phytosanitary Agreement: Issues for Developing Countries in UNCTAD", *Positive Agenda for Future Trade Negotiations*. New York and Geneva.

ANNEXURE A-5.1

UNCTAD Classification of Environment Related Trade Barriers

1. Finance Measures

4174 Refundable deposit for sensitive product to protect environment

2. Automatic Licensing Measures

5271 Prior surveillance to protect human wealth

5274 Prior surveillance to protect environment

3. Quantity Control Measures

6171 Authorisation to protect human health

6172 Authorisation to protect animal health

6173 Authorisation to protect plant health

6174 Authorisation to protect environment

6175 Authorisation to protect wildlife

6176 Authorisation to ensure human safety

Quotas for Sensitive Product

6271 Quota to protect human health

6274 Quota to protect environment (Montreal Protocol)

Prohibition

6371 Prohibiton to protect human health

6372 Prohibition to protect animal health and life

6373 Prohibition to protect plant health

6374 Prohibition to protect environment

6375 Prohibition to protect wildlife

6377 Prohibition to ensure human safety

4. Monopolistic Measures

7171 Single channel for imports to protect human health

5. Technical Measures

Product characteristics requirements to protect human health

Product characteristics requirements to protect animal health and life

Product characteristics requirements to protect plant health

Product characteristics requirements to protect environment

Product characteristics requirements to protect wildlife

Product characteristics requirements to ensure human safety

Marketing Requirements

8121 Marketing requirements to protect human health

8123 Marketing requirements to protect plant health

8124 Marketing requirements to protect environment

8127 Marketing requirements to ensure human safety

Labelling Requirements

8131 Labelling requirements to protect human health

8132 Labelling requirements to protect animal health and life

8133 Labelling requirements to protect plant health

8134 Labelling requirements to protect environment

8135 Labelling requirements to protect wildlife

8137 Labelling requirements to ensure human safety

Packaging Requirements

8141 Packaging requirements to protect human health

8142 Packaging requirements to protect animal health and life

8147 Packaging requirements to ensure human safety

Testing, Inspection or Quarantine Requirements

8151 Testing, inspection or quarantine requirements to protect human health

8152 Testing, inspection or quarantine requirements to protect animal health and life

8153 Testing, inspection or quarantine requirements to protect plant health

8154 Testing, inspection or quarantine requirements to protect environment

8155 Testing, inspection or quarantine requirements to protect wildlife

8156 Testing, inspection or quarantine requirements to ensure human safety

ANNEXURE A-5.2

Commodities Susceptible to ETBs in India's Exports

Commodity Code	*Commodity Name*	*Quantity*	*Value*
10410	Sheep	20000	0.2
1051100	Fowls of the Species *Gallus domesticus*	1504112	0.49
1051900	Other	1546831	0.37
10600	Other Live Animals	0	0.01
2011000	Carcasses & Half-carcasses of Bovine Animals, Fresh/Chilled	2330151	2.04
2012000	Other Cuts of Bovine Animals with Bone in, Fresh/Chilled	1152795	0.92
2013000	Boneless of Bovine Animals, Fresh or Chilled	44759811	47.57
2021000	Carcasses & Half-carcasses of Bovine Animals, Frozen	983624	0.87
2022000	Other Cut Meat of Bovine Animals, Frozen with Bone in	1078110	1.08
2023000	Boneless of Bovine Animals, Frozen	103651656	111.8
2031900	Other Meat of Swine, Fresh or Chilled	33800	0.04
2032100	Carcasses and Half-carcasses, Frozen	25058	0.03
2032900	Other Meat of Swine, Frozen	48540	0.05
2041000	Carcasses & Half-carcasses of Lamb, Fresh or Chilled	1437235	2.67
2042200	Other Cuts with Bone of Sheep, Fresh or Chilled	909116	1.62
2042300	Boneless Meat of Sheep, Fresh or Chilled	684559	1.44
2043000	Carcasses & Half-carcasses of Lamb, Frozen	22400	0.04
2044100	Carcasses & Half-carcasses of Sheep, Frozen	180865	0.36
2044200	Other Cuts of Sheep with Bone in, Frozen	3081608	7.9

Contd. ...

...Contd. ...

Commodity Code	*Commodity Name*	*Quantity*	*Value*
2044300	Boneless Meat of Sheep, Frozen	637422	1.7
20690	Other Edible Offals, Frozen	77960	0.08
2090000	Pig Fat Free of Lean Meat & Poultry Fat (Not Rendered), Fresh, Chilled, Frozen, Salted, in Brine, Dried or Smoked	36536	0.09
21090	Other, Including Edible Flours & Meals of Meat or Meat Offal	10856	0.14
3011000	Live Ornamental Fish	19839	0.59
3019900	Other Live Fish	12028	0.05
3021900	Other Salmonidae Excluding Livers and Roes Fresh or Chilled	14347	0.03
30269	Other Fish Fresh or Chilled Excluding Livers & Roes	2690455	5.1
3032900	Other *Salmonidae*, Excluding Livers & Roes Frozen	5000	0.02
3034900	Other Tunas Excluding Livers & Roes Frozen	5209	0.01
30379	Other Frozen Fish, Excluding Livers & Roes	0	135.09
3041000	Fish Fillets & Other Fish Meat (Whether or Not Minced) Fresh or Chilled	227598	0.3
30420	Frozen Fillets (Whether or Not Minced) Frozen	4504685	8.02
3049000	Other Fillets (Excl. Frozen) & Other Fish Meat	2092227	3.53
3053000	Fish Fillets Dried Salted/in Brin Not Smoked	366317	0.36
3054900	Other Smoked Fish Including Fillets	719832	0.84
30559	Other Dried Fish Whether or Not Salted Not Smoked	3180907	2.6
30613	Shrimps & Prawns Frozen	118072722	738.03
30799	Other Molluscs Excluding Live Fresh/Chilled	374320	0.76
40210	In Powder, Granules or Other Solid Forms, of a Fat Content, by Weight, Not Exceeding 1.5%.	793214	1.2
4022100	Milk & Cream in Pwdr., grnls./Other Solid Forms Not Containing Added Sugar/Other Swetng. Matter Containing Fat Exceedng 1.5% by wt.	487	0

Contd...

...*Contd.* ...

Commodity Code	*Commodity Name*	*Quantity*	*Value*
40299	Other	2019	0
40390	Other	15100	0.02
4049000	Products Other than Whey Consisting of Natural Milk Constituents	40390	0.05
4062000	Grated or Powdered Cheese of All Kinds	680	0
4063000	Processed Cheese not Grated/Powdered	1200	0
4069000	Other Cheese	33822	0.11
40700	Birds' Eggs, in Shell, Fresh. Preserved or Cooked	10884909	11.86
41000	Edible Products of Animal Origin, not Elsewhere Specified or Included	230	0
50400	Guts, Bladders & Stomachs of Animals (Other Than Fish), Whole and Pieces Thereof	382056	3.23
50590	Other	65852	0.49
50690	Other	1599906	0.55
50800	Coral and Similar Materials, Unworked or Simply Prepared but Not Otherwise Worked; Shell of Molluscs, Crustaceans or Echinoderms & Cuttle-bone, Unworked or Simply Prepared but Not Cut to Shape, Powd. UNIT	51676	0.09
51000	Ambergris, Castoreum, Civet and Musk; Cantharides; Bile, (Whether or Not Dried); Glands & Other Animal Products Used in the Preparation of Pharmaceutical Products, Fresh, Chilled, Frozen or Otherwise UNIT	7433	1.02
5111000	Bovine Semen	70	0.01
51191	Products of Fish or Crustaceans, Molluscs or Other Aquatic Invertebrates; Dead Animals of Chapter 3	744876	6.99
51199	Other.	2521	0.08
6011000	Bulbs, Tubers, Tuberous Roots, Corms, Crown & Rhizomes, Dormant	1589679	0.16

Contd. ...

...Contd. ...

Commodity Code	Commodity Name	Quantity	Value
60120	Bulbs, Tubers, Tuberous Roots, Corms, Crowns & Rhizomes, in Growth or in Flower; Chicory Plants & Roots	299962	0.1
6021000	Unrooted Cuttings & Slips of Live Plants	361070	0.12
60220	Trees, Shrubs and Bushes, Grafted or Not of Kinds which Bear Edible Fruit or Nuts	2440	0.01
6031000	Cut flwrs. & Flower Buds Suitable for Bouquets or Ornamental Purposes, Fresh	2722348	5.97
6039000	Other Cut Flwrs. & Flower Buds Suitable for Bouquets/for Ornmntl. Purposes	8295433	12.13
6049900	Foliage Branchs etc., not Fresh Without Flwr./Flwr. Buds & Grasses Suitable for Bouquets/Ornamental Purposes Excl. Fresh	3072788	4.32
7011000	Potato Seeds Fresh or Chilled	389096	0.1
7019000	Potatoes Fresh or Chilled Other than Seeds	7873325	1.23
7020000	Tomatoes, Fresh or Chilled	642712	0.11
70310	Onions and Shallots	215765607	41.85
7032000	Garlic Fresh or Chilled	3592247	0.97
70990	Other	38819734	15.11
7109000	Mixture of Vegetables Frozen	3288584	1.85
71290	Other Vegetables; Mixtures of Vegetables	683622	0.89
7131000	Peas (*Pisum sativum*) Dried & Shld.	303652	0.13
7133100	Beans of the Spp Vigna Mungo, Hepper or Vigna Radiata, Wilczek Dried & Shld.	9880	0.01
71339	Other	18010	0.01
8021100	Almonds Fresh or Dried in Shell	268	0
8021200	Shelled Almonds Fresh or Dried	20412	0.08
8051000	Oranges Fresh or Dried	10407098	2.43
8061000	Grapes Fresh	11382177	8.81

Contd. ...

...Contd. ...

Commodity Code	*Commodity Name*	*Quantity*	*Value*
80620	Dried	143334	0.19
8081000	Apples Fresh	7442120	2.38
8082000	Pears & Quinces Fresh	181894	0.02
81090	Other	12229068	4.99
90111	Coffee Neither Roasted Nor Decaffeinated	178945149	329.7
10011000	Durum Wheat	4100	0
100190	Other	1756900	0.32
10051000	Maize Seed	417160	0.13
10059000	Other Maize (Corn)	1646193	0.32
100630	Semi-milled or Wholly Milled Rice, Whether or Not Polished or Glazed	4888064962	1477.14
10070000	Grain Sorghum	22000	0
11010000	Wheat or Meslin Flour	3955265	0.86
120220	Shelled, Whether or Not Broken	51877132	29.92
12060000	Sunflower Seeds, Whether or Not Broken	1739754	1.65
120799	Other	21089622	11.75
12092900	Other Seeds of Forage Plants	1189190	0.5
120991	Vegetable Seeds	3773368	8.73
120999	Other	1063222	6.03
121190	Other	37976307	56.51
130190	Other	6599663	78.02
140490	Other	5080607	4.82
150410	Fish-liver Oils & Their Fractions	23684	0.19

Contd. ...

...Contd. ...

Commodity Code	*Commodity Name*	*Quantity*	*Value*
16023900	Other Prpd./Preserved Poultry of Hdng. No 0105	201620	0.52
16024900	Other Prpd./Preserved Parts of Swine Incl. Mixtures	71	0
16025000	Prpd./Preserved Meat, Meat Offal of Bovine Animals	40000	0.02
160300	Extracts and Juices of Meat, Fish or Crustaceans, Molluscs or Other Aquatic Invertebrates	154060	0.56
16041900	Other Fish, Whl./Pcs, not Minced, Prpd./Preserved	3663	0
16042000	Other Prepared or Preserved Fish	163564	0.8
190110	Preparations for Infant Use, Put Up For Retail Sale	3878800	6.76
190190	Other	8275108	12.6
19021100	Uncooked Pasta Ntstfd./Otherwise Prpd Containing Eggs	17610	0.03
19021900	Other Uncooked Notstfd./Otherwise Prpd. Pasta	80880	0.1
19022000	Stuffed Pasta Whether or Not Cooked/Otherwise Prpd.	62427	0.08
19023000	Other Pasta	36752	0.06
190410	Prepared Foods Obtained by the Swelling or Roasting of Cereals or Cereal Products	2277631	1.65
19049000	Cereals in Grain form Excl. Maize Pre-ckd, etc.	686309	0.73
190530	Sweet Biscuits; Waffles & Wafers	3007072	3.23
19054000	Rusks Toasted Bread & Similar Toasted Prdcts	90629	0.16
19059000	Others	487812	0.84
21041000	Soups, Broths & Preparations	40550	0.02
21042000	Homogenised Composite Food Preparations	8942	0.03
230120	Flours, Meals & Pellets, of Fish or of Crustaceans, Molluscs or Other Aquatic Invertebrates	112275	0.06
230990	Other	83130500	15.34
300110	Glands & Other Organs, Dried, Whether or not Powdered	61766	1.39

Contd. ...

...Contd. ...

Commodity Code	Commodity Name	Quantity	Value
300120	Extracts of Glands or Other Organs or of their Secretions	39797	0.66
30019000	Heparin Salts Other Human/Animals Substns. Fbr. Thruptc./Prophylactic Use nes	29604	0.46
300210	Antisera & Other Blood Fractions	10336	0.45
300220	Vaccines for Human Medicine	655058	25.04
300290	Other	0	2.75
300310	Mdcmnts. Containing Pnclns./Thr. Drvtvs. with A/Pnclnc. Acd Strctre., Strptmcns./Thr. Drvts.	119557	0.76
300320	Medicaments Cont. other Anti-biotics	33508	0.34
300339	Other Medicaments	523952	5.68
300390	Other	14004768	126.41
300410	Containing Penicillins or Derivatives Thereof, with a Penicillanic Acid Structure,. or Streptomycins or their Derivatives	4012017	72.06
300420	Containing other Antibiotics.	2695801	81.26
300431	Containing Insulin	16811	2.81
41039000	Raw Hides & Skins of Other Animal	2000	0.03
	Total	**5970847060**	**3576.79**

Source: *India Trades*, 2001.

ANNEXURE A-5.3

Selected List of Disputes Referring to SPS and TBT Agreements in WTO DSU

	Petitioner	*Respondent*	*Product*	*Issue*	*Outcome*
DS2	Venezuela	United States	Petrochemicals	Environmental	Appellate Rept.
DS3	United States	Korea	Agriculture	Food Safety	Consultation
DS4	Brazil	United States	Petrochemicals	Environmental	Appellate Rept.
DS5	United States	Korea	Agriculture	Food Safety	Settlement
DS12	Peru	EC	Agriculture	Marketing	Settlement
DS14	Chile	EC	Agriculture	Marketing	Settlement
DS18	Canada	Australia	Agriculture	Quantitative	Appellate Rept.
DS20	Canada	Korea	Agriculture	Food Safety	Settlement
DS21	United States	Australia	Agriculture	Quantitative	Active
DS26	United States	EC	Agriculture	Food Safety	Arbitration
DS41	United States	Korea	Agriculture	Food Safety	Consultations
DS48	Canada	EC	Agriculture	Food Safety	Arbitration
DS56	United States	Argentina	Textiles	Customs	Appellate Rept.
DS61	Philippines	United States	Agriculture	Environmental	Consultations
DS72	New Zealand	EC	Agriculture	Customs	Settlement
DS76	United States	Japan	Agriculture	Food Safety	Appellate Rept.
DS85	EC	United States	Textiles	Transit	Settlement

Contd. ...

...*Contd.* ...

	Petitioner	*Respondent*	*Product*	*Issue*	*Outcome*
DS96	EC	India	Mixed	Customs	Settlement
DS100	EC	United States	Agriculture	Food Safety	Consultations
DS133	Switzerland	Slovak Republic	Agriculture	Transit	Consultations
DS134	India	EC	Agriculture	Customs	Consultations
DS135	EC	United States	Textiles	Customs	Active
DS137	Canada	EC	Agriculture	Quantitative	Consultations
DS144	Canada	United States	Agriculture	Transit	Consultations
DS151	EC	United States	Textiles	Customs	Consultations

Source: RIS Discussion Paper # 22.

6

India's Response

SHASHI SAREEN

Introduction

With establishment of the WTO, the international scenario has changed rapidly with opportunities being available to all countries to benefit from greater access to world markets. International trade has increased and this global trade is further expanding rapidly and significantly due to increase in consumer demands linking to education and awareness of consumers, internationalisation of food tastes and habits, development in food science and technology, improvement in transportation and cold chain linkages.

However, India's contribution to world trade is relatively insignificant. One of the major factors for such low exports is the non-tariff measures in terms of standards and conformity assessment procedures being applied by various countries which are acting as non-tariff barriers.

To ensure that standards and regulations do not create unnecessary barriers to trade, certain rules and disciplines are required to be maintained, which are being taken care of through the non-tariff agreements, especially the SPS Agreement. The SPS Agreement stresses on the health and safety aspects and permits member countries to impose measures to protect the health and safety of their population within certain rules. It also refers to use of Codex standards.

In the present scenario, the major issues, which are acting as non-tariff barriers, have been highlighted

Harmonisation

The SPS Agreement provides for harmonisation of SPS measures of member countries with international standards, the Codex, in the case of food products. The SPS Agreement, however, allows Member States to lay down standards more stringent than Codex but these need to have a scientific justification. It has been observed that certain countries are at times laying down norms more stringent than those specified by Codex and without any scientific basis which are difficult to meet; or the test methods specified are different to those specified by Codex and are not validated or have a very high level of sensitivity which may not be justified and due to which the cost of testing becomes disproportionately high.

Transparency

The SPS Agreement requires governments to notify other countries of any new or changed sanitary or phyto-sanitary measures which affect trade. They also allow for a reasonable interval between the publication of such regulations and their entry into force in order to allow time for producers in exporting members and particularly developing country member to adapt their products and methods of production to the requirements of the importing members. To be able to meet the SPS requirements of various WTO member countries, there is a need to have access to the norms laid down by them. It has often been observed that there is absence of information and lack of transparency on the procedural norms and regulations of various countries as related to specifications as well as methods of sampling, inspection and test. New regulations are brought out and implemented without even giving the producers in the exporting country a chance to get familiar with these. Often the standards are available only in the language of the importing country or are presented in a very complicated manner. The result is that exporters are, at times, not clear about the specific requirements prescribed by the country of destination, which has led to rejection at the point of import.

Risk-based Approach

Risk to consumers resulting from hazard in foods has been identified as a significant concern at the international level. Article 5 of the SPS Agreement provides that sanitary and phyto-sanitary measures should be based on risk assessment taking into account an appropriate assessment of the actual risk involved and if requested by the exporting country make known details of this assessment. Codex Alimentarius Commission is also promoting constant application of risk analysis principles throughout all Codex Committees in setting up standards.

It has been observed that some importing developed countries are fixing standards without carrying out any risk assessment work. An example can be cited in the case of marine products where consignments are being rejected due to presence of certain microorganisms such as *Vibrio parahaemolyticus* i.e. a nil limit has been laid down. In this connection, it may be pertinent to bring out that *Vibrio parahaemolyticus* is a habitant of the marine environment of the tropical waters and there is every chance for the presence of this organisms in raw fish and fishery products. They are generally destroyed during chilling/freezing or by heating at 60 degrees C. Besides, the organism is not considered as a potential hazard in raw frozen products which are to be cooked before consumption. Further, generally importing countries such as Japan have specified limits for *Vibrio parahaemolyticus* only for ready-to-eat cooked products or seafoods for raw consumption and not in products which are to be cooked before consumption. Here again limits have been specified at levels ranging from 1000 to 10000 per gram.

It is understood that no such risk evaluation has been carried out by the importing country as inspite of repeated requests the same has not been made available.

Safety Management Systems Approach

In addition to end product criteria, it is increasingly becoming necessary that food items are produced and processed taking into account critical factors to ensure quality and safety throughout the food chain from primary production till final consumption. HACCP, a food safety assurance and management system, has been recognised by the Codex Alimentarius Commission as a tool to assess hazards and establish control systems with focus on preventive measures instead of relying primarily on end product testing for health and safety aspects. The HACCP-based approach is currently being incorporated into the new hygienic codes under development by the Codex Commission. Such a safety management systems approach is also being insisted upon by many countries including USA, Canada, Australia and many members of the EU both in the domestic and overseas trade presently for more risk prone products such as marine, meat, poultry and dairy.

Some countries are insisting on freezing the infrastructural requirements which is not called for in the HACCP standards and becomes difficult to implement and in any case it is internationally accepted that alternate equivalent measures are permitted if these meet the requirements of the importing country. Another significant issue is implementation of such systems at the primary production level, which is being initiated in certain sectors such as marine and egg products.

Capacity Building

To comply with the requirements of developed importing countries, capacity building is important. Article 9 of the SPS Agreement provides for extending technical assistance to developing country members to enable them to comply with the SPS measures needed to achieve the appropriate level of SPS protection in markets of the importing countries.

Some of the areas identified for capacity building by India to meet requirements of importing countries especially of the EU include, upgrading testing facilities (very important today in the area of residue analysis); upgrading or empowering human resources in areas of testing, risk analysis, development and auditing of HACCP plans etc; development of GMP/GHP/HACCP modules for implementation at national level as well as for exports; establishing databases on requirements of importing countries. Such assistance has so far been very minimal. Although importing countries are coming forward to offer assistance, this is either coming too late by which time the exporting country has already lost out on its markets or has spent significant resources of its own to upgrade its facilities.

Equivalence

The concept of equivalence has been recognised in the SPS Agreement and is also being encouraged at the international level by the Codex Alimentarius Commission with a view to using pooled resources more effectively, avoiding duplication of inspection and testing, and ensuring that health and safety requirements are met effectively.

Government of India, is emphasising on development of Equivalence Agreements with the health authorities of major trading partners. Export Inspection Council (EIC), as the Official Certifying Body of the Government of India, has already been designated as Competent Authority by the European Commission for marine products and *Basmati* rice and is awaiting recognition for egg products and milk products. EIC has also finalised an Equivalence Agreement with AQIS, Australia and the Agreements with Sri Lanka and Italy are in an advanced stage. EIC has also entered into dialogue for the purpose with USA, Canada, Argentina, Japan, South Korea and other countries including those of the EU. Under such agreements, in addition to recognition of our certification, it is proposed to cover exchange of information on standards; methods of sampling, inspection and test, and also to have a provision of retest and appeal in case of rejections, return of rejected consignments, etc. These would serve as an important means of facilitating trade by recognition of our standards and our certification systems to provide for an equivalent level of protection against health risks as those of the importing countries and also lead to reduced rejection rates and provide for reduced inspection of our products in overseas markets.

Such agreements are being encouraged under both the SPS Agreement and Codex standards, and the SPS Agreement states that, "Members shall upon request, enter into consultation with the aim of achieving bilateral or multilateral agreements or recognition of the equivalence of specified SPS measures."

However, members often avoid such agreements even after receipt of a formal request as either the administrative burden of entering into these is not justified or they do not want to lose their control over imports. Some countries have even expressed that import controls is a means of their income and by signing such agreements they would lose financially and therefore are not interested to enter into such 'Agreements'. At times it is seen that important components are not addressed in such agreements as these are not in the interest of the importing country.

Rejection and Destruction of Consignments

Certain health authorities have recently started destroying the contaminated consignments (either because of biological or chemical contamination) instead of returning them to the exporting countries as requested by the exporters/importers. Further, they are not even providing full details of the methods of sampling, analysis and results obtained.

The document on guidelines for exchange of information between countries on rejection of imported food developed by the Codex Committee on Food Import and Export Inspection and Certification Systems (CCFICS), lays down that in case of rejections, the importing country should provide all details of methods of sampling, analysis and results obtained. Further, as per the standard, the importing country can take a decision on destruction of the consignment. In such cases the exporting country is not being consulted. It is necessary to involve the exporting country in such decisions for various reasons as given below:

a) The assessment of brought back consignments has, at times, shown absence of contamination. It is, therefore felt necessary that the consignments found contaminated in the importing country are brought back to enable the competent authority to retest them and ascertain whether the consignments were contaminated or not and if contaminated examine the cause.

b) It has been observed that on number of occasions, the importing country is adopting different methods for sampling and testing and also testing for parameters/contaminants, which are not notified in their standards, which at time become reasons for rejections e.g. countries like Italy are rejecting the consignments testing positive for all strains including non-pathogenic ones for *Vibrio cholerae*, whereas only the '01' strain is pathogenic. Destruction of such consignments may not be justified. Eggs are also there of consignments which have tested positive in one lab of the importing country and negative in another lab. Therefore unless there is uniformity in methods of testing being followed, such rejections would not be justified.

c) By conducting complete re-testing of the brought back consignments, it would be possible to identify the contaminants if any and take immediate corrective measures to control/eliminate its recurrence.

d) Destruction of a consignment leads to wastage of a large amount of money and some cases of contamination can be taken care of through reprocessing.

Health Certificates

Language Problem

Health certificates for export consignments of food products are issued by the competent authority in a prescribed proforma as per the requirements of the importing countries. Certain health authorities have since some time started insisting that the health certificates be filled in their local language for e.g. Spain is requiring that the health certificate for consignments meant for export into their country be filled in Spanish instead of in English. Filling of health certificates in a foreign language becomes difficult and such issues are becoming NTBs.

Retrospective Issue

Another example is in the case of issuance of health certificates retrospectively. While as a rule health certificates are issued before shipment of the consignments, but at times, due to unavoidable circumstances, the exporter may not be able to apply for the health certificate on time and the same may need to be issued retrospectively. Some countries do not accept health certificates issued retrospectively and they insist that health certificates are issued by the competent authority prior to shipment of the consignments.

In this connection, it may be mentioned that the above Codex guidelines do provide for issuance of certificates on retrospective basis if agreed by the competent authorities of the importing and exporting country. The clause in the Codex guidelines is as follows:

"Certificates may be issued while consignments are in transit to the country of destination only when appropriate systems of control are agreed by the competent authorities of the importing and exporting countries."

It is important that such certificates might, at times, be allowed retrospective issue, especially for consignments from exporters approved by the competent authority and who are under the regular surveillance system.

Economic Impact of Certain Measures

Voluntary Standards

Imposition of voluntary international standards such as ISO 14000 on Environmental Management Systems (EMS) by buyers on their suppliers in exporting countries has the effect of not only restricting market access for at least sometime until the industry upgrades itself, but also leading to high cost of implementation. The standard on Social Accountability, SA 8000 is a recently announced international standard for management systems primarily dealing with working conditions, offering better quality of life to workmen and other socio-economic issues. Under the guise of SA, the importing countries can limit the imports of various products on alleged violation of any of the above parameters.

Rapid Alert System

Rapid alert is imposed by a member state of EC in case a consignment does not meet the SPS requirement. When rapid alert is imposed by one Member State of EC, the implementation or effect of this is in all EU countries. For lifting the alert, a specific number of consignments have to be cleared by the country that had imposed the alert. The fact that a large number of consignments had been tested and cleared at the ports of other EU countries has no relevance. Implementation of such a procedure results in 'on alert' at times continuing over hundreds of consignments and in one case over a period of two years. This has very large economic impact. As 'on alert' is imposed on rejection of a consignment at a single port and implemented by all countries, similarly, it should be lifted when a total of 10 consignments to various EU Member States are cleared.

Conformity Assessment Issues

There have been cases of rejections due to use of test methods which are different to those specified internationally that is by Codex. An example can be cited in the case of chloramphenicol in shrimps where the international method is by HPLC, which has sensitivity at a level of parts per million whereas the European Commission is using HPLC/MS/MS which has a sensitivity to the level of part per billion. This very high level of sensitivity may not be justified as the cost of testing becomes disproportionately high. It has also been observed that the method used by Norway for *Vibrio cholerae* is a non-validated method and is different to the Codex method, which has resulted in consignments showing different test results when tested in their laboratory as compared to the results obtained from Indian laboratories.

Due to differing test methods, rejection of consignments is increasing, this issue needs to be looked into. One solution would be to provide for joint testing with involvement of the competent authority. Such provisions are available with some developed countries. Equivalence agreements would also help minimise such issues as these would result in acceptance of the certification system of the exporting country.

Conclusions

Emergence of WTO regime has led to dismantling of barriers—tariff and non-tariff—quality and safety issues have become important which has given an impetus to adoption of international standards by member countries both for product quality as well as conformity assessment procedures, thereby protecting consumers and preventing unfair trade practices while at the same time ensuring free flow of trade across borders. Importing countries are at times imposing conditions that are not in the spirit of the WTO agreements. India needs to take advantage of these agreements and international standards for addressing various concerns either by taking up with the importing governments or through amendments in the international standards especially Codex standards. Member countries also need to actively take up non-tariff issues at various fora including the WTO. This would go a long way in ensuring that non-tariff barriers are not imposed in an unjustifiable manner and the presence of goods from developing countries is increased in global markets.

7

Case Study of Pakistan

SHAHEEN RAFI KHAN
TOQEER AHMAD

Introduction

The relationship between trade and environment has always generated heated debate. Proponents of free trade or globalisation assert the compatibility of trade and environmental objectives. They contend that the most pressing environmental problems will be addressed only when economies are developed sufficiently—trade being one among the many instruments for such development (for instance, through clean technology transfers).[1] The opposing view holds trade/globalisation responsible for degrading the environment and rules out the possibility of synchronising trade and environment objectives. This debate also coincides with a North-South split, one complicated further by the endorsement of Southern concerns by progressive Northern NGOs. Perhaps environmental regulations define this split best. Its advocates in the North advance the view that such regulations will promote sustainable development and growth eventually. Inverting this argument, Southern countries claim environmental regulations are trade-restricting devices, engineered by coalitions to protect domestic industries and block exports from the South.

This chapter tries to avoid the histrionics such debates can give rise to, and focuses on the functional implications. It begins with the premise that the reality is somewhere in between, that trade can have both environmentally malign and beneficial impacts. By the same token, environmental regulations can cut both ways: they can be trade

1. For a good treatment of the various effects, see Munasinghe (1996).

restricting but they also offer new market niches and can lead to cleaner production practices in the exporting countries. The institutional challenge is to address the concerns and maximise the benefits. Also, in the final analysis, developing countries are left with little choice other than to comply with the increasingly stringent environmental regulations in order to maintain their export shares. Given the multi-stakeholder interests, this means satisfying the demands of a range of stakeholders—governments, environmental organisations and the corporate sector.

The North, too, can facilitate this process by displaying sensitivity for institutional and social realities in the South. With regard to process standards (effluents, emissions) tolerances vary, given the initial pollution and emission baselines. This is as true intra-north, as across the North-South divide. Also, compliance/enforcement should be a function of both awareness and institutional capability and the National Environmental Quality Standards (NEQS) should reflect these differences. Such capability embraces in-house environmental management, as well as external (government, delegated) facilities to measure, monitor, audit and enforce. In particular, spillover and downstream impacts, which make attribution difficult inherently underscore the need for precise measurement. Effective mechanisms are as critical when assessing environmental/health effects embodied in products. Failure to adhere to prescribed standards—by passing off products through spurious certification—can erode credibility and affect adversely export prospects.

Trade-Environment Linkages in Pakistan

Over the past few years, the textile sector has undergone changes in term of composition and the increased export of untraditional but value added items (Table 7.1). Due to these changes, the textile sector is now able to produce wrinkle-free, anti-bacterial, anti-dust and anti-cling fabric that a few years ago was unknown. Similarly, it now possesses the ability to process 320-cm wide cloth with 24 colours. Besides traditional items such as yarn and cotton fabrics, the textile sector now offers a wide range of quality products that include bed covers, bed sheets, bed skirts, bed ruffle, bedspread, waterbed sets, quilts covers, rod valance, table covers, comforters, curtains, pillow covers, pillow sham and pillow ticking. Other new products being

Table 7.1

Pakistan's Share in World Export

(US$ Million)

	1993	*1994*	*1995*
Total Trade	3,257,589	6,842	0.21
World Total	3,617,591	7,328	0.20
Pakistan (% Share)	4,490,031	8,125	0.18
Textile & Clothing	271,926	238,433	277,669
World Total	5,091	5,593	5,921
Pakistan (% Share)	2.23	2.35	2.13
Carpets & Rugs	1,440	1,213	1,090
World Total	158	187	149
Pakistan (% Share)	10.97	15.42	13.67
Leather Goods	12,697	14,645	18,041
World Total	240	269	280
Pakistan (% Share)	1.89	1.84	1.55
Fish Products	33,384	36,531	41,693
World Total	202	143	150
Pakistan % Share	0.61	0.39	0.36

Source: Ahmad, 1998.

introduced high quality satins for beds and furniture and twills for apparel and clothing.

This sector illustrates the diversity of linkages between trade and environment. It embraces the gamut of product, technology, scale and structural effects. The environmental impacts have a two-fold dimension. Negative product effects are evident in the use of substandard and unbranded chemicals and dyestuffs, particularly azo-dyes. Obsolete technology in bleaching and singeing operations wastes energy and water. Scale effects manifest themselves in the increasing discharge of wastewater, contaminating water and soil and threatening aquatic life. The adverse structural effects relate to the growth of textile processing over the past two decades with its attendant demand for fresh water, the best transferring agent in processing. In the Kharianwala area of Faisalabad (the hub of textile industries), the underground water level has decreased from 10-15 feet to 200-250 feet as a result of excessive pumping of ground water by these newly established textile-processing units.

But there are positive environmental offsets as well. Benign product effects are embodied in the reduction of carbon content in fabrics. Alternative dyestuffs are replacing increasingly toxic dyes such as sulphur black or azo-dyes. The increasing scale of operations has made possible the introduction of a range of the latest environment friendly technology. In operations like printing and dyeing, energy efficient machinery and equipment imported from Germany and Italy are in use. The introduction of thermosole[2] dyeing machines has generated economic efficiency as well as ecological gains. The increasing use of pad batch dyeing for cotton, rayon and blends conserves energy, water, dyes and chemicals, labour and floor space. Water consumption for pad batch dyeing with steam wash-off is typically under two gallons per pound of dyed fabric, compared to, typically, 20 or more on atmospheric beck for the same fibre reactive dyed shade. This reduces associated BOD and COD loading for waste streams by up to 80 per cent (Bailey, 1998).

The Financial, Social and Institutional Implications of Compliance

Though no empirical study has been done which quantifies the financial costs and benefits of environmental compliance, some inferences are possible. The export figures (Table 7.1) show a steady rise in Pakistan's exports but its share in world trade is declining particularly of textiles, which are the mainstay of its exports. This trend is reflected in dollar terms (Table 7.1). This is discouraging, considering the subsidies to exporters through rebates and the frequent rupee devaluations.

While declining shares cannot be attributed solely to the lack of environmental compliance, the domestic policy climate is one which does not allow much scope for exploiting emerging market niches. The export companies in Pakistan have opted for an out-moded premise of international trade, i.e. selling low-cost, low quality items, cushioned by domestic protectionist policies. But such policies are becoming increasingly costly. The ban by the US on the import of surgical goods from Pakistan in 1994 due to lack of quality certification and an attempt to ban the export of footballs in 1997 signal the importance of familiarising industrialists with standard requirements (Ahmed, 1997).

2. Thermosole dyeing machines is considered better than traditionally used jiggers and jets.

Protectionist policies, being inward-looking, encourage poor corporate behaviour. Fortunately, some firms have shown foresight, reaping gains by upgrading their production technologies. However, while this does have beneficial environmental spin-offs, these are not enough. For instance, the thermosole dyeing machine provides an example of environmental benefits accruing through increased efficiency. But it needs to be combined with end-of-pipe solutions to reduce the increased outflow of untreated wastewater. Further benefits can accrue from improved environmental management. But, this is easier said than done. It is an area where compliance strategies and policies have to be given careful consideration.

Social conditions, environmental tolerances and institutional capabilities set a cap on how far developing countries are prepared to go in accommodating Northern standards. This is as true in the North as in the South. Countries with a high environmental profile, like the Netherlands, Denmark and Sweden developed their own systems of environmental management, reflecting their own internal realities and requirements. As a consequence standards are not harmonised intra-North, which establishes a precedent for the South to chart its own route to compliance.

The Environmental Standards

Various health and environmental hazards are associated with product use and its disposal, sometimes termed as "product ecology." Product oriented environmental standards have emerged out of scientific research. Such research has revealed the presence of chemical substances in various industrial products injurious to human health. This has led to the promulgation of environmental laws and regulations in developed countries. These laws regulate not only the composition and characteristics of products but also mandate safe disposal after use. They are primarily Northern-driven with direct implications for the South, as they prohibit the entry of any product, which does not comply with these laws.

In environmentally sensitised countries like Sweden, Denmark and Holland, concern is growing about the harmful effects of textile fabrics. It has evoked reactions on issues like azo-dyes and formaldehydes. In Sweden, the campaign for clean cloth is aimed at educating buyers and consumers about the harmful effects embodied in the fabric used. The Danish Environmental Protection Act (EPA) has published lists of colorants (dyestuffs), which are frequent skin sensitisers and should be regarded as carcinogens. Germany has banned the use of azo-dyes altogether and prohibits any product containing the trace of azo-dye from entering the German market.[3]

These refer to the negative effects of production and process methods on the local environment and on the health of people. Over the past few years, concerted efforts have been made to harmonise process and production methods worldwide. The standards, as embodied in various eco-labelling schemes and environmental management systems (ISO-14000), aim to reduce the effluent and emission levels generated by production processes. In developing countries, people most affected are industrial labour and poor communities residing adjacent to industrial sites. The premise is that environmental awareness among such people is low. They are often ignorant of the health risks of using chemicals for instance, and have little opportunity to influence their working environment and immediate surroundings.

For instance, a study conducted in 1992 indicates that in the five ginneries in rural Sindh, 9 per cent of total workers had severe byssionosis, 64 per cent were asymptotic and the rest had other respiratory symptoms (Khan, *et al.*, 1999). A survey conducted in tannery cluster of Sialkot assessed the environmental and health impacts. A very high proportion of communities, over 85 per cent, held the tanneries responsible for the majority of public health and environmental problems. While the majority favoured shifting the tanneries out of their localities, a substantial minority voted to retain them, provided they instituted appropriate waste management and environmental protection measures (Ahmed, 2000).

The North favours harmonised production and process methods by arguing that the transition to environment friendly production

3. Any product with azo-dyes that may split off any one of the 20 carcinogenic amines is not prohibited in these countries. The splitting of azo-dye can occur on the skin, in the intestine and in the body. On the skin and in the intestine, this is brought about by bacteria. In the body, for instance in the liver, azo compounds can be broken down by certain enzyme systems. Non-fixed, water soluble azo-dyes can also come in contact with skin *via* perspiration fluid. Some of these azo-dyes form amines, which are carcinogenic. Benzidine, in particular, is an identified carcinogen.

processes in developing nations would be an extremely important factor in improving both the working and living environments, and with it the general health of the local population. Professing equal concern, Southern governments resist harmonisation for the reasons cited above. However, converging objectives improve prospects for negotiated agreements.

Spillover effects can be both regional and global, in as much as production activities do not recognise local or national boundaries. Acid rain deposition and the discharge of effluents into river bodies, with adverse downstream impacts on agriculture and biodiversity, are examples of regional effects. Global warming and ozone layer depletion exemplify global impacts. The effects can be of the quick-response variety (industrial effluents, acid rain) or may entail long lead times (global warming, ozone depletion). Transnational agreements are required to deal with the cross-border dimension of environmental problems (Kyoto and Montreal Protocols, Agreements on POPs).

Environmental standards and related developments in the North, which affect, primarily, the textile and leather products of Pakistan have been divided into three major categories. Category 1 deals with environmental laws, which regulate the nature and composition of importing products. Category 2 focusses on eco-labelling requirements of various European countries. Category 3 looks at process-oriented instruments, which are usually a part of environmental management systems.

Environmental laws relate to product-based environmental standards. Among complying Pakistani export firms, the capability of dealing with these bans is variable. While exporters have, by and large, been able to restrict the use of azo-dyes, the presence of metals like cadmium asbestos and nickel is more difficult to detect. Companies either do not have access to the required detection facilities, or they are not available. Details of various environmental laws and regulations, which can affect the export potential of major trade products, have been provided in Annexures A-7.3 and A-7.4.

The experience with various European-based eco-labelling schemes has also been mixed. The elements of each scheme are relatively uniform across European countries, which is an advantage, and are based on the criteria of production ecology.[4] However, some schemes are more stringent than others in their requirements.

For instance, as per one scheme (the Nordic Eco-labelling Scheme) it is possible to meet the condition of not using carcinogenic, mutagenic, toxic or potentially sensitising dyes. This applies, too, to limitations on extractable materials. Less amenable is the requirement of 90 per cent biodegradability of detergents and other chemicals used in different processes; limits on discharge of wastewater; and limits on AOX emissions. Such changes require expensive technology upgrades or emission control devices. Similarly, the ban on tetrachlorophenol (TCP) & pentachlorophenol (PCP) and formaldehydes practically prohibit the use of local chemicals and compel the producers to use environment friendly chemicals[5] from more reliable sources, i.e. multinational companies. This is expensive and hikes up the product price quite considerably.

However, other eco-labelling schemes based upon product standards only, are easier to comply with. The Oeko-Tex standard-100 points is one such scheme.[6]

The standard-100[7] sets yardsticks for noxious substances. It puts limits on the concentration of substances which, according to present knowledge, could have dangerous effects on humans during normal use. Oeko-Tex belongs to a product based standard series and is independent of manufacturing processes, the only requirement being that certain harmful substances not be used. Using this label fulfills the normal eco-label requirements. Pakistani textile export firms have reacted positively to it and concerted efforts are underway to qualify for this label. For more details on European eco-labelling schemes and Nordic eco-labelling schemes, see Annexures A-7.5, A-7.6 and A-7.7.

4. This refers to the process of production and manufacture of fibres, textiles and garments which should be environment friendly, and should satisfy requirements for air emissions, effluents, waste treatment and protection against noise.

5. Some Western companies have developed alternatives like Busan-30, which are 30 times costlier than PCP or equivalent chemicals (Kaushik, 1999).

6. Few companies like Chenab Fabrics have already qualified for it.

7. The Oeko-Tex Standard-100 is based upon the criterion for human ecology, which in the view of consumers is more significant. This is based on the effect of garments on the users and on their near surroundings.

The third category relates to the development and implementation of environmental management systems. Given the prevailing discrepancies in implementing national environmental standards, the international buyers consider compliance with international standards such as ISO 14001 as the best way to guarantee the fulfillment of the supplier's environmental obligations, particularly in relation to local environmental laws.[8] A brief comparison of three popularly known environmental management system (ISO 14000 series, Environmental Management & Audit Schemes and BS 7750) is given in Annexure A-7.8.

Environmental management systems (ISO 14001 and EMAS) requires a commitment to comply with the environmental performance requirement that already exist in laws, regulations and voluntary guidelines and ensures that that implementing companies pursue a policy of 'continual improvements'. Under EMS framework performance of the organisation is measured with respect to management of aspects and impacts on the basis of measurable results of the environmental management system, related to an organisation's control of its environmental aspects, based on its environmental policy, objectives and targets.

Local firms face barrier from the certification and verification systems and complain about the high costs associated with being certified to ISO 14001.[9] Many SMEs experience insufficient drivers for adoption and are uncertain about the market benefits of such systems. The lack of sector specific guidance and material tailored to different sizes of firms is an added barrier.

The lack of human resources rather than financial ones is the major internal barrier to EMS implementation and becomes increasingly important as the size of the company decreases. Practical problems with EMS implementation about how to determine environmental aspects and assign significance and how to achieve internal auditor independence in firms also exist. SMEs are largely ill-informed about environmental management systems, how they work and what benefits can be gained from their implementation.

8. The ISO 14000 series is modeled after the BS 7750 (Environmental Management Systems) originally published in 1992 and updated in 1994.
9. BS 7750 has been withdrawn in favour of ISO 14000 while EMAS are essentially for companies operating in European markets.

The Environmental Impacts

Pakistan has two main hubs of industrial development: Karachi, with its proximate industrial estates and industrial corridors in the Punjab. Both are characterised by their lack of waste management facilities, giving rise to acute environmental and health problems. With the majority of industries, untreated water is discharged directly into the main drain; flowing eventually into river bodies and seeping into underground aquifers and contaminating them. The pollution in Ravi River is the highest among all rivers in the Punjab. At present, the wastewater flow from the city of Lahore is 19 m 3/sec with a BOD load of about 346 tonnes/day. Industries around Lahore discharge heavy metals (Cu, Ng, Ni, Cd, Cr) into the river water, with an estimated total load of 2.6 tonnes/day. The quantities of heavy metals (Cr, Cu, Pb) discharged are estimated at 15 tonnes/day (Khan, 2000).

Heavy metals, like chromium, are carcinogenic. High BOD and COD loads deplete oxygen level in fresh water bodies, causing mortality of aquatic life. The presence of heavy suspended solids also causes this. In addition, it blocks penetration of sunrays into lower strata of water bodies restricting their ability to perform photosynthesis, a process necessary for their survival (Box 7.1).

Groundwater in the cities of Faisalabad (due to textile dyeing mills), and Karachi (due to diversified industrial operations) has also become contaminated. Industries which cannot access drainage channels, *nullahs* or rivers, resort to land disposal of their effluents. The wastewater accumulates in depressions. Some industries use soakage pits to dispose off their effluents. This leads to groundwater contamination.

Textile Industry

By any measure, textiles are the most important sector of Pakistan's economy whether it be production, employment generation, consumption and exports (Annexure A-7.2). In recent years, environment-based restrictions have been added to the more conventional trade restrictions. The most important of these pertain to the ban on the use of azo-dyes, which possess carcinogenic properties (Banuri, 1998).

BOX 7.1

The Case of Kasur City

Kasur has a population of about 0.27 million and is located about 55 km southeast of Lahore. With its 240 tanneries, Kasur has the largest concentration of tanneries in the country. Currently only 9 per cent treat their effluent. In 80 per cent of the cases, the effluent is discharged untreated into recipient water bodies or onto open land. In only 11 per cent of the cases; is the wastewater drained into municipal sewers. Along with untreated municipal sewage, it is discharged into the Rohi *Nullah* flowing eventually into the Sutlej River.

The tannery clusters produce around 13,000 m3/d of effluents containing high BOD, TDS and toxic chemicals such as chromium, sulfides, salt and dyes. Farmers along the Rohi *Nullah* use the mixed sewage for irrigation. As reported by them, the crop yields have decreased by 50 per cent, perhaps due to high concentration of sodium. Water supplies from hand pumps are contaminated with dissolved solids, especially sodium chloride and sulfides. Sulphur present in the polluted water from hand-pumps is affecting metallic utensils. Improper use and handling of chemicals has subjected workers and residents to hazardous exposure. Respiratory disorders, skin infection related diseases, diarrhoea/dysentery are the most serious illnesses among the community (Azher, 2000).

Environmental impacts in the cotton chain manifest themselves early, beginning with the use of pesticides and fertilisers. In Pakistan, evidence exists that pests such as the American bollworm and the whitefly have developed resistance against common pesticides leading to progressively larger doses, a condition referred to as the pesticides treadmill.

Various textile processes, using chemicals and dyestuffs, cause environmental and health problems of different kinds. These problems may occur during the production process (emissions, effluents, waste disposal, occupational or downstream health impacts), or the effects may be caused by exposure to the product itself, for instance, skin allergies. Table 7.2 provides details of the links between the various stages of production and their environmental and health impacts.

Eco-labelling has become an increasingly important tool for discerning consumers in industrial countries to influence production

Table 7.2

Environmental Hazards Associated with Textile Sector

Process	*Source*	*Major Impacts*
Fibre Cultivation	Pesticides use	Occupational health problems, reduction in natural fertility for the soil, harm to soil structure, soil aeration and soil erosion & reduced genetic biodiversity.
Spinning	Spinning operation	Occupational health problems causing acute respiratory disease, low biodiversity & smog.
Washing	Detergents, soaps, alkalis, wetting agents, foamers, defoamers and lubricants with high BOD, COD, TDS	Carcinogenic, depletion of ozone layer, potentially bioaccumulative, obnoxious odour, water turbidity, reduce light penetration, threaten aquatic life & low biodegradability.
Sizing/Desizing	Sizing material	Low biodegradability.
Bleaching	Bleaching material	Low biodegradability, emission of air pollutants such as formaldehydes etc.
Mercerizing		High pH value.
Dyeing	Dyestuff & chemicals	Carcinogenic, high allergy risk, high BOD & COD, obnoxious odour, threatens fish spawning grounds & water system corrosion.
Finishing	Finishing materials	Extremely persistent effluents, very low biodegradability, high toxicity & occupational health hazards.

Source: Based on information from various sources cited above.

Table 7.3

Environmental Codes of Conduct

Company	*Standards for Suppliers*
Wal-Mart	Environment: In accordance with the principle of the three 'Rs'—reduce, reuse and recycle.
Sears & Roebuck	Strict adherence to local laws governing working conditions and production methods.
Levi Strauss	Sharing commitment to the environment and conducting business in a way that is consistent with Levi Strauss & Co.'s environmental philosophy and guiding principles.
IKEA	Due compliance to environmental conditions such as use of azo-free dyes, formaldehyde, PCP (pentachloropehenol).
C & A	C & A will assist its suppliers in their efforts to meet their obligation to preserve the environment.
Hennes & Mauritz	Suppliers must comply with all environmental laws and regulations in the country of operation.

processes. It is also being encouraged by various major marketing chains (Table 7.3). The transmission works both ways: eco-labelling schemes provide information to consumers and encourage producers to shift to more environment-friendly production processes. Environmental labels contain information about the environmental impacts and qualities of the products, enabling organisations and individuals to make informed choices. Eco-labelling also constitutes a market-based incentive, both in terms of the choices it provides to consumers and the niche opportunities it offers to producers.

These schemes, originally designed for trade among developed countries, are being extended increasingly to products from developing countries. One such example is Century Textiles of Bombay, the largest textile company in India. When its products gained Öko-Tex certification, it was able to raise prices by 8-10 per cent; simultaneously, the market grew by 10 per cent (SIDA, 1998). In addition, a scope for potential 'new' markets is also being established. Synthetic materials can be replaced by natural ones, for instance, jute and sisal are offered as substitutes for artificial fibres, and colour grown cotton can replace chemical intensive artificially dyed cotton.

Table 7.4

Environmental Hazards Associated with Leather Sector

Waste	*Source*	*Major Impacts*
Liquid	Organic & inorganic solid, hydrogen sulphide, ammonia, volatile organic compound, electrolytes.	Occupational health problems, corrosion of the water-carrying system, contamination of coastal area, reduction in natural fertility for the soil, harm to soil structure, threat to aquatic life, low crop yield, contamination of food, and reduction in genetic biodiversity
Solid Waste	Organic material, meat remains, the tanned skin remains, chromium remains during trimming.	Contamination of soil and underground water, contamination of food, occupational health hazard such as perforation and bronchiogenic carcinoma, contamination of poultry feed, reduce genetic biodiversity.
Air Emissions	Hydrogen sulphide & sulphur-containing compound, ammonia & nitrogen containing compounds, organic solvents.	Lower than NEQS.

Source: Based on information from different sources.

An Italian organisation, namely, Legambiente, is promoting an international event called Ecomoda for the fifth year: its aim is to spread information about ecological solutions for the production of textiles.[10] All the major producers will present their products and proposals. The European Commission is supporting the initiative, as a lot of attention will focus on the European eco-label.

Leather Industry

On a unit basis, leather tanning is ranked among the highest for its pollution impacts. Chemical inputs are used intensively at every stage, from the cleaning of hides in the slaughterhouse, to their processing upto the final stage.

The solid wastes contain chromium residues, which are known to cause perforation and bronchiogenic carcinoma in humans who are continuously exposed (Table 7.4). Chicken feed prepared from proteins containing tannery wastes introduces chromium directly into the food chain. Tests conducted by the Pakistan Tannery Association showed chrome residues in poultry feed. Leather shavings used as cheap fuel in kilns also release chromium into the environment. Solid wastes are dumped around the factory area on unutilised lands. These contain metals such as chromium, aluminum and zirconium, with detrimental effects on plant growth (Khan, *et al.*, 1999).

As in the case of textile effluents, large pH fluctuations and high BOD concentrations kill all natural life in affected water-bodies. Informal surveys have revealed that tannery wastes have polluted the Korangi (Karachi) and Charsadda (near Peshawer) areas of Pakistan. The contribution of tanneries to the contamination of the Karachi coastal waters is estimated at about 10-15 per cent of the total. Hydrogen sulphide formed due to the presence of sulphide in the effluent and chromium proved so toxic that, workers in Karachi died while clearing monsoon ditches filled with tannery sludge. In Punjab, tanneries are contaminating directly prime agriculture land. The informal surveys also show that that the crop-yield has been affected adversely and the food contaminated (Ahmad, 2000). Most of the tanneries in the Punjab and North-West Frontier Provinces (NWFP) are located in or close to residential neighborhoods and pose a serious threat to the urban population.

10. Source: *http://europa.eu.int/comm/environment/ecolabel/envir.htm.*

Environmental Sensitivity of Industries

The environmental sensitivity of the industrial products/sectors included in this study (which could also be used as a selection criteria to identify products/sectors for future research) rests on four aspects. The first aspect relates to the extent of degradation of natural resources and biodiversity loss. The second criteria reflects pollution control expenditures—or potential pollution charges in case of non-compliance. The third aspect accounts for current and potential export trends. The fourth aspect takes into account employment sensitivity and employment generation potential.

The Environmental Standard Committee (ESC) of Pakistan has classified industries into three main categories in order to establish monitoring and reporting guidelines for industrial effluents and gaseous emissions. The legal basis for such classification is the Environmental Protection Bill, 1997, Section 6 (1) (i), which authorises the Federal Agency to "establish systems and procedures for surveys, surveillance, monitoring, measurement, examination, investigation, research, inspection and audit to prevent and control pollution, and to estimate the costs of cleaning up pollution and rehabilitating the environment in various sectors." These monitoring and reporting guidelines are based on environmental sensitivity of the industries. The following three factors have been taken into account while developing the classification system:

- pollution levels,
- toxicity of pollutants, and
- cost of monitoring.

This classification system is designed basically to assess environmental impacts qualitatively, as no comprehensive data is available on industry/sector based pollution loads or air emissions.[11]

Category A refers to the most hazardous industries which include textile processing industry, tanning & leather industry, petroleum refining, fertilisers, chemicals. Category B refers to moderately hazardous industries, such as dairy industry, fruit and vegetable

11. Industries are listed in two different lists for liquid and gaseous emissions. One industrial unit can be in Category A for liquid emission parameters and in Category B for gaseous emission.

processing, sugar, detergent, etc. In Category C are the least hazardous industries, such as pharmaceuticals, marble, cement industries.[12]

The pollution charges have been calculated on the basis of liquid discharge from an average industrial unit (Table 7.5). Only two pollutants, COD and TSS, have been taken into consideration to calculate annual charges to be paid by these industrial units.[13] The base pollution charge will be Rs. 50 during the years 1999 and 2000. The charge will increase incrementally. The charges will be incremental in nature as given below:

Table 7.5

Financial Implications of Pollution Charges

Pollution Charges (Million Rs./Year)					
Industry	*Year 1*	*Year 2*	*Year 3*	*Year 4*	*Year 5*
Sugar	0.457	0.915	2.747	4.121	5.495
Paper	1.745	3.491	10.475	15.713	20.950
Textile	0.142	0.285	0.856	1.285	1.713
Leather[14]	1.709	3.419	10.258	15.388	20.517
Pollution Charges (% of Sale)					
Industry	*Year 1*	*Year 2*	*Year 3*	*Year 4*	*Year 5*
Sugar	0.05	0.09	0.27	0.41	0.55
Paper	0.34	0.69	2.05	3.08	4.11
Textile	0.01	0.02	0.06	0.09	0.12

01 Jan. 1999 to 31 Dec. 1999: 10 per cent increase on base rate

01 Jan. 2000 to 31 Dec. 2000: 20 per cent increase on base rate

01 Jan. 2001 to 31 Dec. 2001: 40 per cent increase on base rate

01 Jan. 2002 to 31 Dec. 2002: 60 per cent increase on base rate

01 Jan. 2003 to 31 Dec. 2003: 80 per cent increase on base rate

12. "Guidelines for Self-monitoring and Reporting by the Industry", SDPI 1998.

13. For details, see Annex 3.

14. The data for leather industry is from "Leather Sector: Environmental Report", ETPI, *http://www.etpi.org.pk/pdf/LeatherSectorReport.pdf 1998.*

A brief review of current export trend is provided in Annexure A-7.2 while the data for daily average employment in each sector covers those establishments, which are engaged in manufacturing activities under the Factories Act, 1934. The total reported employment during 1990-91 was 622,234. The largest number of persons were engaged by textile industry which employed 268,494. Industry-wise details are given in Table 7.6.

Table 7.6

Daily Average Employment by Industry

Industries	*Average Daily Person Engaged*	*Average Daily Employment Inc. Contract Labour*
Textile	268494	268121
Leather	9390	9359
Footwear	5254	5241
Sports Goods	7720	7705
Surgical Instruments	3999	3970
Fish	NA	NA

Source: *Census of Manufacturing Industries 1990-91*, Federal Bureau of Statistics, Government of Pakistan.

Based on the selection criteria above, the textile and leather industries are clearly the most sensitive environmentally. Table 7.7 provides a qualitative interpretation of the mix of quantitative and qualitative data. The three industries/sectors scoring highest by the composite criteria are cotton textiles, leather and woolen carpets.

Table 7.7

Ranking Environmental Sensitivity of Industries/Products

Components	*Sector*					
	Textile	*Leather*	*Carpets*	*Sports Goods*	*Surgical Instruments*	*Fish*
Environmental Impacts	5	5	4	1	1	2
Pollution Charges	4	5	4	0	1	0
Export Potential	5	2	3	3	4	2
Employment	5	3	2	2	2	NA
Total	19	15	13	6	9	4
Ranking by Importance	1	2	3	5	4	6

Note: Score (5=very high, 4=high, 3=middle, 2=low, 1=very low, 0=none)

Source: *Census of Manufacturing Industries 1990-91*, Federal Bureau of Statistics, Government of Pakistan.

Mitigating Initiatives

Legislation aimed at environment protection was introduced for the first time in 1977. Since then many institutional, policy and regulatory developments have taken place at the federal and the provincial levels. These include, among others, the creation of a Ministry of Environment, federal and provincial level environmental protection agencies, promulgation of the Pakistan Environmental Protection Ordinance in 1983, development of the Pakistan National Conservation Strategy (NCS) in 1992 and promulgation of the Pakistan Environmental Protection Act in 1997. Also, during this period, several protection, conservation and abatement projects have been implemented.

A substantive breakthrough in environmental protection followed the first meeting of the Pakistan Environmental Protection Council (PEPC) in 1993, which led to the approval of the NEQS for industrial emissions and effluents. The NEQS can be regarded as the first step towards internalising external effects. They consist of 32 liquid and 16 gaseous parameters,[15] in addition to setting limits on noise pollution. Industry specific NEQS and standards for water and land receiving bodies have been developed recently but are yet to be notified by the Ministry of Environment. Different standards have been set for effluents going into inland water bodies, sewage treatment plants, and into the sea. For instance, effluents discharged into the municipal sewerage system can have higher pollutant loads, provided they are processed eventually in a treatment plant. Changes have also been made in the standards of gaseous emissions.

A pollution charge regime (see details above) is being introduced to achieve industrial compliance with the NEQS.[16] These standards, if successfully implemented and documented, would go a long way in meeting the standard requirements likely to be imposed by importing countries (Khan, 2000). The aim of the charge is to discourage environmentally damaging activities and strengthen the incentives to

15. See ETPI, "Environmental Report on Leather Sector", *Environmental News*, January 2001.

16. A pollution charge is a fee or tax on amount of pollution in excess of levels allowed by the NEQS

reduce waste and pollution while generating revenues, which will be earmarked for environmental protection.[17] The principle is that charge should induce clean up operations in industries. In other words, the net cost of the clean up should be less than the pollution charge.

This development is unprecedented in Pakistan as the industrial sector has agreed voluntarily to a charge to be applied for generating pollution in excess of permissible national limits.

The objectives of the NEQS are unlikely to be achieved without a proper and systematic environmental monitoring programme. The Pakistan Environmental Protection Council (PEPC) approved the 'Guidelines for Self-Monitoring and Reporting by the Industry' on August 26, 1999. The supporting guidelines for self-monitoring and reporting by industry include details on the sampling procedures, handling, transport, storage and preservation of samples, procedures for analyses of various pollutants and their flow rate measurements. Under these guidelines, all industries in Pakistan are required legally to submit self-monitoring reports from January 1, 2000.

Normally the monitoring is carried out by government agencies. However, the innovative aspect of this programme is that industries decided to take the responsibility of monitoring and reporting of industrial effluents and emission upon themselves. The Federal EPA, in close association with the provincial EPAs, has initiated the first phase of a voluntary pilot programme and nearly 50 industrial units have participated in the programme. The national EPA has agreed also to issue a green certificate to the first 100 industries.

Other areas are being identified to develop consensus-based policies between environmental monitoring institutions and the industrial associations, based on shared scientific and financial data. One joint initiative currently underway is that between the Pakistan Sugar Manufacturers Association (PSMA), the Pakistan Society of Sugar Technologists (PSST), and the Pakistan Paper & Pulp Board Manufacturers Association (PPPBMA). The associations have prepared already the first draft of the 'Environmental Policy' of their respective

17. The secular aspect of these pollution charges is that money collected will be used primarily for activities that will help in abatement of environmental pollution through soft loan and other incentives. The private sector has demanded the creation of Provincial Environmental Trust Funds (PETFs) for that purpose.

sectors. The draft of the Sugar Sector Environmental Policy was prepared in active consultation with PSMA and PSST. In the paper sector, the major players, i.e Century Paper & Board Mills Ltd. and Packages Ltd., agreed to share their substantial experience in the implementation of environmental solutions with The Environmental Technology Programme for Industries (ETPI), for preparation of the policy paper for their sector. Eventually, these documents will be presented to the federal and provincial EPAs.

Private Initiatives

Reflecting a combination of international pressure and national environmental legislation, various industrial associations in the country have begun to take intermediate and long term actions towards making the industrial sector environmentally benign.

A number of interesting projects are underway. Among these are the; 'Environmental Management Programme of Pakistan Tanners Association', 'The Introduction of Cleaner Technologies in Punjab Tanneries', and 'The Uplift Programme for Korangi Industrial Area'.

The Korangi Association of Trade and Industry is the largest industrial estate of the country. About 2400 industrial units are located in the estate. The association has developed a 15-year phased clean-up programme for itself. In the first phase, cleaner production options will be promoted and implemented. In the second phase, small sector level combined treatment plants will be installed. Finally in the third phase, a combined effluent plant will be installed for NEQS compliance.

The Pakistan Tanners Association (PTA) has been implementing a comprehensive environmental programme since 1992. This includes a combined effluent treatment plant, a solid waste management programme, an occupational health and safety programme, and a drainage system for about 160 tanneries located in Korangi, Sector 7-A. The total cost of the project is more than Rs. one billion. The Sindh Government's, Export Development Fund, and individual tanners are providing more than 95 per cent of the finances.

In the Punjab, the PTA is implementing Cleaner Technology Programme with the assistance of The Netherlands Government. The project has completed an initial environmental examination of more than 260 tanneries. Prototype environmental packages are under implementation in the selected tanneries. The emphasis of the project is on in-house improvements, primary treatment systems, and chemical recovery plants. The project will be completed by the year 2000.

Technology Transfer for Sustainable Industrial Developments (TTSID)

The Technology Transfer for Sustainable Industrial Developments (TTSID) is one of the important initiatives in the family of private sector environmental projects. Funded by the Swiss Federal Office for Foreign Economic Affairs, and housed at SDPI, the project provided support for the promotion of sustainable industrial production through business-government roundtables, technical research reports, and information packages on environmental issue. The training component of the TTSID developed training materials by conducting environmental studies, followed by hands-on training and workshops. The project also conducted a study to establish green credit facilities for industrial environmental management projects.

Environmental Technology Programme for Industries (ETPI)

The Environmental Technology Programme for Industries is an independent initiative currently under way to extend technical services to industry and the government. Its primary objective is to promote the use of environmentally safe technologies in the industrial sector. This five-year project began in 1996 and works with Pakistani industries and their associations in identifying the most economical pollution prevention and abatement technologies and in implementing these solutions. The five components of the programme include: the development of a user-friendly database of relevant information; institutional networking within and between key industrial institutions of the country; dissemination and communication to promote cleaner industrial production; institutional support and training to create in-house environmental capacity within chambers and industrial associations; and demonstration projects in 20 selected industrial sub-sectors to demonstrate the economic feasibility and environmental efficacy of environmental technologies. Three representative industrial units have been selected in each sub-sector for preliminary environmental audits to assess the extent and nature of the environmental problems.

The ETPI is a joint project of the Federation of Pakistan Chambers of Commerce and Industry (FPCCI) and the Government of the Netherlands. Considered a private sector project, its equity base is 67:33, with the 33 per cent grant provided by the Government of Netherlands.

Cleaner Production

Another private sector initiative is cleaner production through optimisation of resources and utilisation of waste on commercial basis. (Boxes 7.1, 7.2). It aims to promote the use of waste from different sectors as useful by-products for other industries. For example, in tanneries fleshing and trimmings of wet-blue leather ordinarily cannot be used due to chrome content. These are now being treated for use as chicken feed, leather board and glue manufacturing. In paper mills the recovered fibre is used for manufacturing of paperboard. Textile solid wastes are converted to waste cloth for use in other industries.

BOX 7.2

Chrome Recovery Plants–The Environmental Income

From 25-30 per cent of the chrome that goes into the wastewater can be reused. The chrome tanning effluent from the tanning drum is the influent to the chrome recovery system. Addition of sufficient alkali to spent chrome recover will cause precipitation of the chromium in the form of basic complexes. These are separated from the liquor by settling and/or filtering under pressure. The liquor is almost free of chromium and contains most of the dissolved solids and other impurities. The chrome sludge cake can be dissolved in sulfuric acid to form tanning liquor which can then be reused. Preliminary estimates of the economic viability of chrome recovery and reuse in tanneries suggest that the cost recovery can be achieved in 6 to 7 months for most tanneries. This assumes 95 per cent chrome recovery, which is very much within the feasible range. The capital cost of such a plant, for a tannery processing, around 12,000 kg of hides per day, is around US$ 20,000.

Source: Environmental News, September, 1999.

BOX 7.3

Shakarganj Sugar Mills Limited–A Zero Effluent Industry

Shakarganj Sugar Mills Limited is located in district Jhang, Central Punjab, with an installed sugar production capacity of 10,000 tonnes per day. The main reasons for sugar losses were, overloading of the evaporators, vacuum pans, excessive boiling and incorrect liquid levels. The stream represents about 92 per cent of the wastewater from Shakarganj Mills. The presence of BOD and COD in the spray pond overflow was due mainly to the presence of sugar content in the wastewater. The pollution load was very high, with most of it being organic in nature.

The mill had already constructed lagoons for wastewater treatment. These were not operating at the desired efficiency levels, due to flotating of fly ash and oil on the surface of water, which hindered the biological decomposition processes. ETPI conducted the environmental audit and recommended that ashes and oil be trapped at source. Sand filter and oil skimmers were installed, with an immediate and substantial increase in efficiency. ETPI proposed a three weeks composting method based on the mixing of mud cake with the distillery wastewater. The idea has been successfully tested on 25 per cent of the wastewater from the distillery. The compost is being used successfully in the sugarcane farms. At present, a full-scale plant is under construction. ETPI also recommended a system of environmental monitoring and evaporators maintenance. After the implementation of the proposed system the pollution levels are within the limits of National Environmental Quality Standards

The Shakarganj Mills invested in the range of US$50,000 to US $75,000. In comparison, end-of-pipe solutions would have cost US$1.25 million.

Source: Environmental News, September, 1999.

Research is underway on the use of the recovered lignin from the black liquor as a binding material in different industrial products. Manufacturing of compost by mixing the mud cakes with the distillery waste water is one of the best examples of making economically productive by-products from the waste.

Policy Options for Pakistan

In the Netherlands, codes of conduct for clothing vendors require a safe working environment for labour, including attention to safety and health, throughout the production chain, and participation in monitoring these conditions. Moreover, the seller helps finance the compliance system. Until 1996, no vendor signed the code of conduct (CREM, 1996) Swedish companies such as Hennes & Mauritz, KappAhl, IKEA and OBS place both environmental and social demands on their suppliers as a condition of doing business. These demands are not as rigorous as for eco-labelling, but they are mandatory in nature and monitored closely by the buyers.

IKEA, which is currently buying textile products from a number of textile companies in Pakistan, does not permit the use of chemicals such as azo-dyes and cadmium in textile products. The use of formaldehyde in textile products is governed by Finnish regulations. IKEA also applies the strictest possible (German) regulations on PCP (pentachlorophenol), used as a mould agent. These specify a maximum limit of 5-mg PCP/kg, which amounts to a ban on its use effectively.

IKEA also provides a niche for 'organically grown' cotton, meaning that no artificial fertilisers, or chemical biocides have been used in its cultivation.

In many OECD member countries, green public purchasing (GPP) programmes promote government purchases of environment-friendly products. Such practices are gaining currency in many European countries. In Sweden, tenders to supply public authorities are subject to rigorous environmental standards. Through the European Union's Generalised System of Preferences, the European Commission has proposed the establishment of an 'Encouragement Regime'. Import tariffs may be reduced by 20-30 per cent for products produced in 'an environmental and human way' (CREM, 1996). Not surprisingly, exporters from developing countries may view these as schemes that appear to favour domestic producers. However, such concerns may also reflect an inability to compete.

Globalisation and trade liberalisation have both been instrumental in raising environmental concerns worldwide. In fact, the North has become increasingly sensitive to health, environmental, fair trade and labour aspects, embodied in traded products. The challenge is to integrate trade and environmental policies harmoniously in such a way that maximum synergies are achieved. In other words, the ideal paradigm is one where trade policies become environmentally sensitive and environmental policies are not trade-restrictive. A key requirement is that the South be encouraged and assisted in every possible way to take advantage of emerging green markets. While the South has its own environmental agenda, which coincides with many Northern environmental concerns, the task is to ensure their convergence at the institutional level.

Initiatives in the South

Innovative approaches have been applied both by the governments in third world countries and by the private sector to provide green shades to industrial activities and products. Governments now realise that they can increase the influence of communities and stock markets by compiling reliable pollution statistics and reporting them to the public in easily understood forms.

Grading Polluters by Colour

In Indonesia, the idea of grading polluters by colour has proved very effective. The government has cut industrial water pollution sharply by monitoring industrial discharges and publicising their finding in the news media. The government officials rank water polluters using colour codes: black for factories, which make no attempt to control pollution and causing serious damage; red for those which have some pollution control but fall short of compliance; blue for those which meet national standards and green for those which are much cleaner than required.

Environmental News in the Stock Market

In the Philippines, share prices for beer-maker San Miguel shot up nearly 60 per cent when environmental regulators highlighted the firm's 'green' record and the installation of new pollution control equipment. In Mexico, the share prices for paper manufacturer Kimberly Clark fell nearly 50 per cent after the government levied substantial fines for violating water pollution regulations.

The National Agenda: A Checklist of Possible Initiatives

Many environmental initiatives have been undertaken in Pakistan. At present, there tends to be a co-incidental rather than trade-specific aspect to these regulations. A concerted effort needs to be made to integrate industrial and trade policies with environmental objectives. However, environmental measures should be designed keeping in mind the capacity of industries to conform with international environmental standards and the viability of the legal and institutional mechanisms to ensure compliance with such standards. An indicative check-list of themes for further trade-environment policy development is presented below. Essentially, these are areas of national concern, where some have a greater trade orientation than others.

Sector Specific Policies

A possible way to strengthen trade-environment linkages could be through a sector orientation. The case of sugar and pulp & paper industries was mentioned in the previous chapter. Extensive consultations between stake-holders in these sector/industry policy formulations have established an encouraging precedent for incorporating environmental concerns in the export-oriented sectors, such as leather, textiles, fisheries, etc.

Designation of Specific Inputs

Chemicals are used extensively in many industrial processes; in particular, the use of azo-dyes in textile and leather goods processing is common. About 200-300 of the 2000 or so azo-dyes available have carcinogenic amines and are banned in western countries.[18] Compliance with product-embodied standards with respect to azo-dyes could, notionally, entail the following:

- A ban on azo-dyes as listed in Annex I.
- Standardisation of dye manufacturing and other industrial chemicals.
- Provision of the colour index (CI) number and generic name.
- A written proof that azo-dyes will not split off any of the harmful carcinogenic amines.
- A material data safety sheet providing information about the carcinogenic nature and occupational health hazards Backward Linkages.

18. Any product with azo-dyes that may split off any one of the 20 carcinogenic amines (Annex I) is not prohibited in these countries.

Textile exporters cannot prevent the use of harmful pesticides in upstream cultivation. This is where national level policies are required. Adverse trade effects need to be factored in the process where bans on the use of harmful pesticides are being contemplated for health reasons. Although only one pesticide, i.e. PCP is prohibited in product legislation of the EU countries, the number of pesticides subject to prohibition in one or more eco-labelling schemes are increasing. A list of harmful pesticides provided by CREM can be taken as a benchmark. A ban on the use of these pesticides will enhance the capacity of manufacturers to comply with specified standards.

Waste Management

The NEQS and the associated pollution charge regime is expected to generate a thrust towards end-of-pipe treatment options both for air emissions and effluents. Studies conducted through TTSID and ETPI programmes have shown that treatment options are cost effective and that, in additional, a potential exists to convert these options into business opportunities. (Boxes 7.1 and 7.2).

Also measures such as reduction in custom duty on purchase of pollution abatement equipment, income tax relief for using these equipment and most favoured tax treatment to indigenous pollution control technology, etc. are needed to offset the lethargic inertia of the inaction. The inclusion of a comprehensive package for air emission and waste water treatment plants in environment policy will provide the much-needed boost to industries.

Product Packaging

Legislation on packaging and packaging waste is already in force or in the process of being developed in all EU countries. Basically, this applies to the reuse and recycling of packaging material, as well as to limits on toxic concentrations in the material itself. It is an area domestic producers need to start paying more attention to.

The European Union Directive on Packaging and Reuse has introduced strict limits on concentrations of lead, cadmium, mercury, and hexavalent chromium in packaging, which is:

600 ppm after 30 June 1998

250 ppm after 30 June 1999

100 ppm after 30 June 2001

The packaging waste recovery limits vary from 50 per cent to 70 per cent. Compliance saves the importer the costs of composting or incineration.

Industries can ease into compliance gradually. A maximum concentration of 1500 ppm for all heavy metals is suggested, followed by a 10 per cent to 15 per cent reduction annually, until it goes down to some reasonable level. Paper and cardboard packaging, as compared to plastic, glass, wood or metal entails less hazards, therefore should be given preferential treatment. Also, a certain weight percentage of packaging material should be capable of being recycled, composted or incinerated. The use of polyvinyl chloride (PVC) should be avoided in plastic packaging. Finally, standardised multiple-use packaging is a preferred option as opposed to single-use packaging.

References

Ahmad, M. (1997). "Trade & Environment in Pakistan", *Environmental News*, August.

————. (1998). "Implementing ISO 14000 in Pakistan—The Industry's Perspective", *Environmental News*, December.

————. (2000). "Participatory Decision Making for Sustainable Industrial Development: The Case of the Stranded Communities in the Tannery Clusters in Sialkot, Pakistan", *Environmental News*, August.

APTPMA. (1998). *Environmental Newsletter*, Karachi, Pakistan.

Banuri, T. (1998). *Pakistan: Environmental Impact of Cotton Production and Trade*, SDPI.

Bailey, P. (1998). *Pollution Prevention and In-Plant Control Measures in Textile Mills*, SDPI.

CREM (1996). *Analysis of the Opportunities for and Threats to Indian Export as a Result of the Environmental Policy and Trends in the European Union*, The Netherlands.

EPB (1997). *Pakistan Textile Exports Quota and Non-Quota Guide*, Export Promotion Bureau, Karachi, Pakistan.

Khan, Azher Uddin (2000). "Institutional Politics of Environment", *www.cpp.org.pk*

Khan, Haroon Ayub (2000). *A Review of Provincial and District Conservation Strategies*, NCSMTR study prepared for MELGRD, March.

Khan, S.R., M.A. Khwaja, A.M. Khan, H. Ghani, and S. Kazmi (1999). *Environmental Impacts and Mitigation Costs Associated with Cloth and Leather Exports in Pakistan*, SDPI.

Kaushik, A. (1999). *Promoting Sustainable Trade: The Case of Environmental Requirement*, IISD Publications.

Munasinghe, Mohan (1996). "Issues and Policy" *Working Paper No.* 16, Environment Department, The World Bank, Washington, DC.

Richardson, J. (2000). *Trade and Environmental Standards*, Institute of Development Studies, UK.

SIDA (1998). *Report on Trade, Development and Cooperation, www.sida.se/Sida/articles/1600-1699/1630/ter.pdf*, SIDA.

UNCTAD (1999). *Trade and Development Report*, Geneva.

UNEP (1999). *Trade Liberalisation and the Environment*, Geneva.

Annexure A-7.1

Pakistan's Principal Textile Export

(% Share)

Product	1995-96	1994-95	1993-94	1992-93	1991-92
Cotton Yarn	17.6	18.78	18.51	16.47	16.99
Cotton Fabrics	14.6	13.28	12.07	12.67	11.86
Made-ups	8.90	7.96	8	9.04	7.73
Hosiery	8.07	8.47	7.48	6.81	6.16
Garments	7.45	7.89	9	9.07	8.88
Raw Cotton	5.82	0.76	1.16	3.98	7.50
Synthetic Textile	5.25	7.07	9.53	7.38	6.05

Source: EPB, 1997.

Annexure A-7.2

Share of Major Export Products in Pakistan (Rs. Million)

Product	1995-96	1996-97	1997-98	Average
Textile (Excl. Carpets)				
Yarn	54061.911	58491.014	55107.982	55886.969
Cotton Fabric Woven	43278.977	49354.342	53991.139	48874.819
Manmade Fabric Woven	15435.611	20048.744	26729.304	20737.886
Knitted Fabric	2981.578	3339.930	3873.593	3398.367
Other Fabric	2173.907	2084.591	2222.350	2160.283
Bags/Sacks	851.599	1105.812	1079.600	1012.337
Tarpaulin	1322.637	1419.073	2529.459	1757.056
Blankets	230.493	171.588	255.229	219.103
Bed Linen	20487.269	25975.284	31335.435	25932.663
Curtains	838.956	1461.450	1798.835	1366.414
Madeup n.e.s.	3911.629	5115.759	6976.424	5334.604
Apparel & Clothing Accessories	56625.665	69442.872	76722.408	67596.982
Total Textile Export	202200.232	238010.459	262621.758	234277.483
% Share of Total Export	68.60	73.2	70.4	70.7
Leather				
Artificial Leather	0.403	1.215	2.129	1.249
Bovine Skin Leather	1570.074	1966.343	1986.488	1840.968

Contd. ...

... *Contd.* ...

Product	*1995-96*	*1996-97*	*1997-98*	*Average*
Other Bovine Leather	2472.756	2025.872	1934.646	2144.424
Sheep/Lamb Leather	977.422	933.301	628.566	846.429
Goat Leather	3632.699	4363.977	4404.913	4133.863
Leather				
Other Animals	72,509	31,025	13,720	39.084
Furskin	1.463	0	0	0.487
Other Products	245.379	266.250	405.973	305.867
Total Leather Export	8972.705	9587.983	9376.435	9312.374
% Share of Total Export	3.0	2.95	2.51	2.83
Carpets				
Hand Knotted Wool Carpets	6964.382	7577.231	8475.895	7672.502
Hand Woven Rugs	30.490	119.684	116.880	89.018
Other Carpets	136.298	120.462	108.674	121.811
Miscellanous	0.042	3.469	7.779	3.763
Total Carpet Export	7131.212	7820.846	8709.228	7887.095
% Share of Total Export	*2.4*	*2.4*	*2.3*	*2.39*
Surgical Instruments				
Dental Instruments	59.211	101.018	144.196	101.475
Medical Instruments	4230.290	4838.546	5259.648	4776.161
Gas Masks	1.880		4.528	2.136
Medical Furniture	1.773	1.342	2.767	1.960
Miscellaneous	55.208	54.648	328.719	146.191
Total Surgical Instruments Export	4348.362	4995.554	5739.858	5027.924
% Share of Total Export	1.5	1.5	1.5	1.5
Sports Goods				
Footballs	3358.174	3882.167	5057.422	4099.254
Fishing Goods	10.066	1.289	5.472	5.609
Ice Roller & Skates	0.527		1.487	0.671
Snow Skis	2.261		0.750	1.004
Tennis Goods	13.569	3.751	4.078	7.133
Gloves	1938.110	2883.980	3349.392	2723.827
Miscellaneous	3051.837	5359.862	8174.653	5528.784
Total Sports Goods Export	8374.544	12131.049	16593.254	12366.282
% Share of Total Export	2.8	3.7	4.4	3.7

Contd. ...

... Contd. ...

Product	*1995-96*	*1996-97*	*1997-98*	*Average*
Fish				
Fresh Fish	1004.222	1516.953	1548.193	1356.456
Fish Dried	468.315	574.806	526.886	523.336
Crustaceans & Molluscs	3006.330	3566.427	5102.535	3891.764
Others	223.201	140.296	196.364	186.620
Total Fish Export	4702.069	5798.482	7373.978	5958.176
% Share of Total Export	1.6	1.8	1.98	1.78
Rice Basmati	9937.633	7982.238	10967.079	9628.983
Rice Other Varieties	7203.388	10470.725	13595.415	10423.176
Total Rice Export	17141.020	18452.963	24562.494	20052.159
% Share of Total Export	5.8	5.7	6.6	6.06
Total % Share	79.98	85.56	83.19	82.91

Source: SDPI Database 2001.

Annexure A-7.3

Western Environmental Laws on Textile Products

Country	*Products*	*Law*	*Status*	*Standard*	*Parameter*
Germany	Clothing, Clothing Material, Bedding, Towels, Underwear, Outer Clothing, Sportswear	Fourth Act amending the German Commodity Goods Act, July 1995	Legislation	Prohibition	Azo-dyes (Appendix I)
The Netherlands	All Garments, Footwear, Bed Linen	Dutch Commodity Goods Act (Warenwet)	Legislation	Prohibition	Azo-dyes (Appendix I)
Sweden	Clothing, Clothing Material, Bedding, Towels, Underwear, Outer Clothing, Sportswear	Not Available	Future Legislation	Prohibition	Azo-dyes (Appendix I)
Germany	Textile Articles Intended for Skin Contact	Dangerous Substance Act, 1993	Legislation	Compulsory labelling when exceeding 1500 ppm	Formaldehydes
France	Textile Articles Intended for Skin Contact	Not Available	Future Legislation	Not to exceed 200 ppm	Formaldehydes
The Netherlands	Textile Articles Intended for Skin Contact	Not Available	Unofficial Standards	Not to exceed 100 ppm	Formaldehydes

Contd. ...

Contd. ...

Country	*Products*	*Law*	*Status*	*Standard*	*Parameter*
The Netherlands	Textile Products	Staatscourant nr.35, Act, 1994	Legislation	Not to exceed 5 mg/kg.	Tetrachlorophenol
Sweden	Textile Products	Not Available	Future Legislation	Not to exceed 5 mg/kg.	Tetrachlorophenol
Germany	Textile Products Intended for Skin Contact	German Commodity Goods Act, 1992	Legislation	Compulsory labelling when exceeding 0.5 ug/cm2/week.	Nickel & nickel compounds
The Netherlands	Textile Products Intended for Skin Contact	Not Available	Future Legislation	Probably compulsory labelling when exceeding 0.5. ug/cm2/week	Nickel
Sweden	Textile Products Intended for Skin Contact	KIFS 1996. Amendment to the Chemical Products Act, 1985 and KIFS 1992	Legislation	Compulsory labelling when exceeding 0.5 ug/cm2/week.	Nickel
European Union	Textile Articles Intended for Skin Contact	Directive 76/769/ EEC, 1976	Legislation	Prohibition	Flame Retardants
Germany	Textile Articles Intended for Skin Contact	German Commodity Goods Act, 1992	Legislation	Prohibition	Flame Retardants
The Netherlands	Textile Articles Intended for Skin Contact	Textile Articles Decree, 1974	Legislation	Prohibition	Flame Retardants

Contd. ...

... *Contd.* ...

Country	*Products*	*Law*	*Status*	*Standard*	*Parameter*
Sweden	Textile Articles Intended for Skin Contact	Chemical Product Act, 1985	Legislation	Prohibition	Flame Retardants
European Union	Textile Accessories	Directive 76/769/ EEC, 1976	Legislation	Maximum 100 ppm	Cadmium
Germany	Textile Accessories	Chemical Act, 1993	Legislation	Maximum 100 ppm	Cadmium
The Netherlands	Textile Accessories	Staatscourant, 1990	Legislation	Maximum 50 ppm	Cadmium
Sweden	Textile Accessories	Chemical Product Act, 1985	Legislation	Prohibition	Cadmium
European Union	Protective Clothing & Furnishing Textiles	Directive 76/769/ EEC, 1976	Legislation	Prohibition	Asbestos
Germany	Protective Clothing & Furnishing Textiles	Chemical Act, 1993	Legislation	Prohibition	Asbestos
The Netherlands	Protective Clothing & Furnishing Textiles	Staatscourant, 1983	Legislation	Prohibition	Asbestos
Sweden	Protective Clothing & Furnishing Textiles	Chemical Product Act, 1985	Legislation	Prohibition	Asbestos

Source: SDPI Database 2001.

Annexure A-7.4

Western Environmental Standards on Leather Products

Country	*Products*	*Law*	*Status*	*Standard*	*Parameter*
Germany	Leather Products Intended for Skin Contacts that includes Outerwear, Shoes, etc. but Excludes Purse & Other like Products	Fourth Act Amending the German Commodity Goods Act, July 1995	Legislation	Prohibition	Azo-dyes (Appendix I)
The Netherlands	Similar to German Legislation	Dutch Commodity Goods Act (Warenwet)	Future Legislation	Prohibition	Azo-dyes (Appendix I)
Germany	Any Leather Product	Chemical Act, 1993	Legislation	Maximum 5 mg/kg (5 ppm)	Pentachlorophenol
European Union	Any Leather Product	Directive 76/769/ EEC, 1976	Legislation	1000 ppm	Pentachlorophenol
The Netherlands	Any Product	Not Available	Unofficial standards	Maximum 5 ppm	Pentachlorophenol
Germany	Leather Products Dyed by Surface Treatment	Chemical Act, 1993	Legislation	Maximum 100 ppm	Cadmium
The Netherlands	Products using Cadmium as a Stabiliser, Pigment or Coating	Staatscourant, 1990	Legislation	Maximum 50 ppm	Cadmium

Contd. ..

... *Contd.* ...

Country	*Products*	*Law*	*Status*	*Standard*	*Parameter*
European Union	Leather Products	Directive 76/769/ Dyed by any	Legislation EEC, 1976 Treatment Process	Maximum	Cadmium 100 ppm
Germany	Any Product	Chemical Act, 1993	Legislation	Prohibition	Polychlorinated Biphenyles & Eerphenyles
The Netherlands	Any Product	Dangerous Substances Act, 1993	Legislation	Prohibition	Polychlorinated Biphenyles & Eerphenyles
European Union	Any Product	Directive 76/769/ EEC, 1976	Legislation	Prohibition	Polychlorinated Biphenyles & EEC, 1976
European Union	Articles Intendec for Skin Contact	Directive 94/27/ EC, 1994	Legislation	Max. Release of 0.5ug/cm2 per Week of Nicket to Skin	Nickel
Germany	Articles Intended for Skin Contact	German Commodity Goods Act, 1992	Legislation	Max. Release of 0.5 ug/cm2 per Week of Nicket to Skin	Nickel
The Netherlands	Articles Intended for Skin Contact	Not Available	Future Legislation	Max. Release of 0.5 ug/cm2 per Week of Nicket to Skin	Nickel

Source: SDPI Database 2001.

Annexure A-7.5

European Eco-labelling Scheme for Textile Products

Ecological Criteria	*Requirement*	*Compliance Verification*
1. Acrylic	Residual acrylonitrile content < 1.5 mg/kg of yarn. Emission to air of acrylonitrile < 1.5 mg/kg of fibre produced.	Test required on application.
2. Cotton Less than 50% Cotton	Aldrin, captafol, chlordane, DDT, dieldrin, endrin, heptachlor, hexachloroben-zene, hexachlorocyclohexane (total isomers), 2,4,5-T, chlordimeform, chlorobenzilate, dinoseb and its salts, and monocrotophos < 0.05 ppm	Test method: US EPA recommended Test report required on application.
3. Polyster Fibre	Antimony < 300 ppm VOCs emissions < 1.2 g/kg of produced polyster resin	Test method: direct determination by atomic absorption spectrometry.
4. Carding & Spinning Oil, Waxes, Lubricants, Sizes	90% by weight biodegradable.	Test method: ISO & OECD recommended. Test report required on application.
5. Tetrachlorophenol TCP & Pentachlorophenol PCP	Prohibited	
6. Formaldehydes	Prohibited for stripping	
7. Detergents	(a) Alkylphenolethoxylates(APEOs), bis (hydrogenated allowalkyl) dimethylammoniumchloride (DTDMAC), distearyldimethylammoniumchloride (DSDMAC), di (hardenedtallow) dimethyl ammoniumchloride (DHTDMAC) and andethylenediaminetetraacetate (EDTA) are prohibited (b) At each wet-processing site 90% by weight biodegradable.	Test method: ISO & OECD recommended. Test report required on application.

Contd. ...

... *Contd.* ...

Ecological Criteria	*Requirement*	*Compliance Verification*
8. Bleaching Agents	AOX emission < 40 mg C1/kg	Test method: ISO & OECD recommended. Test report required on application only if chlorinated bleaching agents are used.
9. Impurities in Dyes	As < 50 ppm; Cd < 20 ppm; Cr < 100 ppm; Cu < 250ppm; Hg < 4 ppm; Ni < 200 ppm; Pb < 100 ppm; Sb < 50 ppm; Sn < 250 ppm; Zn < 1500ppm	
10. Impurities in Pigments	As < 50 ppm; Cd < 50 ppm; Cr < 100 ppm; Hg < 25 ppm; Pb < 100 ppm; Sb < 250 ppm; Zn < 1000 ppm.	
11. Chrome Mordant Dyeing	Potassium dichromate < 1.8 % Sodium dichromate < 1.5 %	
12. Metal Complex Dyes	< 7 % of the dyestuff applied (input to the process) shall be discharged to waste water treatment (whether on-site or off-site)	Test method: atomic absorption spectrometry. Test report required on application only if chrome mordant dyeing or metal complex dyes are used.
13. Azo-dyes	Prohibited (Appendix 1)	
14. Carcinogenic, Mutagenic or Toxic Dyes	C.I. Solvent Yellow 1; C.I. Solvent Yellow 2; C.I. Solvent Yellow 3; C.I. Basic Red 9; C.I. Disperse Blue 1 and C.I. Acid Red 26 are prohibited. Any dye or dye preparation that may cause cancer, may cause heritable genetic damage, may impair fertility or may cause harm to the unborn child as defined in Council Directive 67/548/EEC, as last amended by Commission Directive 98/73/EEC is prohibited.	Test method: German or French method recommended. Test report required on application.

Contd. ...

... *Contd.* ...

Ecological Criteria	*Requirement*	*Compliance Verification*
15. Potentially Sensitising Dyes	C.I. Disperse Blue 3;C.I. Disperse Blue 35; C.I. Disperse Blue 106; C.I. Disperse Blue 124; C.I. Disperse Yellow 3;C.I. Disperse Orange 3; C.I. Disperse Orange 37/76 and C.I. Disperse Red 1 are prohibited if fastness level is less than 4.	Test method: ISO recommended Test report required on application only if one or more of these dyes are used.
16. Halogenated Carriers	Not allowed	
17. Printing	Printing pastes used shall not contain more than 5 % volatile organic compounds (VOC). Plastisol-based printing is not allowed.	
18. Formaldehyde	< 30 ppm for products intended for infants of less than 2 years of age < 75 ppm for products that come into direct contact with the skin < 300 ppm for all other products.	Test method: Japan Law 112 or Finnish standard. Test report required on application (except for yarns).
19. Waste Water Discharges from Wet-processing	< 25 g/kg of COD content pH between 6 and 9 Temperature < 40 C	Test report and appropriate data required on application.
20. Flame Retardant	Prohibited	

Source: SDPI Database 2001.

Annexure A-7.6

Nordic Eco-Labelling Scheme for Textile Products

Ecological Criteria	*Requirement*	*Compliance Verification*
1. Raw Cotton Fibres	Pesticides are Prohibited	Certification from an expert organisation Test Method: US EPA.
2. Polyster Fibre	Antimony < 300 ppm VOCs emissions < 1.2 g/kg of produced polyster resin	Certification from an expert organisation.
3. Chemicals	A complete list of all chemical products used, including formulas for each Chlorophenyls; PCB (polychlorinated biphenyls); Halogen-based anti-matting agents; Halogen-based moth proofing agents; Chlorine-based bleach; Bromo- and chloro-organic flame-retardants; Organic tin compounds and PVC are prohibited. APEO (alkylphenol ethoxylates); LAS (linear alkylbenzene sulphonates); DADMAC (dialkyl dimethyl ammonium Chloride); Phthalates; EDTA & Halogenated solvents < 1% by weight of chemical product purchased.	Certificates from chemical suppliers (Appendix 3).
4. Spinning Oils and Knitting Oils	Polycyclic aromatic hydrocarbon < 1.0 % by weight	Product information sheet.
5. Azo-dyes	Prohibited (Anex)	Certificate from the dye manufacturer.
6. Adhesive Dressing	At least, inherently biodegradable.	Certificate from the supplier or information sheet.

Contd. ...

... Contd. ...

Ecological Criteria	*Requirement*	*Compliance Verification*
7. Pickling	Prohibited with metals	Certificate from dye works
8. Emission of Oxygen Demanding Substance	For inorganic fibre < 25 g/kg For organic fibre < 60 g/kg	Laboratory report
9. Energy & Water Consumption	A plan for minimising electricity and heat showing how many litres of water and kWh are consumed.	Documentation from the original textile manufacturer.
10. Metals	As < 0.20 mg/kg; Cd < 0.10 mg/kg; Cr < 2.0 mg/kg; Cu < 50.0 mg/kg; Hg < 0.02 mg/kg; Ni < 4.0 mg/kg; Pb < 0.80 mg/kg; Sn < 250 ppm; Zn < 60.0 mg/kg	Laboratory report
11. Formaldehydes Emission from Textile Products	Children clothes < 30 ppm Clothes < 30 ppm Outdoor clothes < 100 ppm Curtains < 300 ppm Furniture fabrics < 300 ppm Carpets < 300 ppm Bed textiles < 30 ppm Other textiles < 30 ppm	Laboratory report
Marketing	Train marketing personnel regarding eco-labelling requirements	Certificate from marketing personnel as shown in Appendix.
Inspection	Conducted by eco-labelling body	
Certification Body	Accredited by the International Federation of Organic Agriculture Movements	Approval by eco-labelling organisation.

Source: SDPI Database 2001.

Annexure A-7.7

European Eco-labelling Scheme for Leather Products

Ecological Criteria	*Requirement*	*Compliance Verification*
1. Residues in the Final Product	Chromium (VI) < 10 ppm Arsenic < 10 ppm Cadmium < 10 ppm Lead < 10 ppm	Test method: Direct determination by atomic absorption spectrometry.
	Free and partially hydrolysable formaldehyde (a) Textile components of footwear < 75 ppm (b) Leather components of footwear < 150 ppm	Textiles: Japan Law Leather: IUC Test report required on application.
2. Emissions from the Production of Material	75 % reduction of the COD contents in waste water.	Test method: ISO & OECD recommended. Test report required on application.
3. Tetrachlorophenol TCP & Pentachlorophenol PCP	Prohibited	
4. Formaldehydes	Prohibited for stripping	
5. Use of Harmful Substances (up until purchase)	(a) Pentachlorphenol (PCP) and its salts and esters shall not be used.	Test method: Textiles: gas chromatography (GC) with electron capture detection (ECD), limit value 0,05 ppm leather: (a) mass spectrometry (MS) or (b) electron capture detection (ECD); limit value 5 ppm.
	(b) No azo-dyes shall be used that may cleave to any of the aromatic amines (Annexure 1).	Textile: German or French method Leather: standard DIN, limit 30 ppm.

Contd. ...

... *Contd.* ...

Ecological Criteria		*Requirement*	*Compliance Verification*
6.	Use of Volatile Organic Compounds (VOCs) During Final Assembly of Shoes	(a) General sports, children, occupational, men's town, specialist cold: 30 gr VOC/pair, (b) Casual, women's town < 25 gr VOC/pair, (c) Fashion, infants, indoor < 20 gr VOC/pair.	Test method: ISO & OECD recommended. Test report required on if chlorinated bleaching agents are used.
7.	Electric Components	The footwear shall not contain any electric or electronic components.	
8.	Packaging of the Final Product	(a) Cardboard boxes shall be made from a minimum of 80% recycled material. (b) Where plastic bags shall be made from recycled material.	
9.	Consumer Information	Potassium dichromate < 1.8% Sodium dichromate < 1.5% < 7% of the dyestuff applied (input to the process) shall be discharged to waste water treatment (whether on-site or off-site);	Test method: atomic absorption spectrometry. Test report required on application only if chrome mordant dyeing or metal complex dyes are used.
10.	User Instructions	The following information shall be supplied with the product: (a) These shoes have been treated to improve their water resistance. They do not require further treatment. (This criterion is applicable only to footwear that has been water-resistant treated.) (b) Where possible repair your footwear rather than throw them away. This is less damaging to the environment.	

Contd. ...

... Contd. ...

	Ecological Criteria	*Requirement*	*Compliance Verification*
11.	Fitness for Use Criteria	(a) Uppers flex resistance (b) Uppers tear strength (c) Uppers bondability (d) Outer soles flex resistance (e) Outer soles abrasion resistance (f) Outer soles bondability (g) Insoles water absorption and desorption (h) Uppers water resistance (i) Outer soles water resistance	Test report required on application.
12.	Printing	Printing pastes used shall not contain more than 5% volatile organic compounds (VOC) Plastisol-based printing is not allowed.	
13.	Formaldehyde	< 30 ppm for products intended for infants of less than 2 years of age.	Test method: Japan Law 112 or Finnish standard.
		< 75 ppm for products that come into direct contact with the skin.	Test report required on application (except for yarns).
		< 300 ppm for all other products.	
14.	Waste Water Discharges from Wet-processing	(a) < 25 g/kg of COD content (b) pH between 6 and 9 (c) Temperature < 40 C	Test report and appropriate data required on application.
15.	Flame Retardant	Prohibited	

Source: SDPI Database 2001.

Annexure A-7.8

Brief Comparison of BS 7750, EMAS and ISO 14000

	BS 7750	*EMAS*	*14001*
Nature	National	Regional	International
Scope	Whole site	Whole site	Part of the site
Environmental Aspect	Examination & assessment of environmental effects	Full inventory of environmental aspects	Identification of significant environmental aspects.
Continuous Improvement	Required	Required	Emphasised
Objectives	Time-scaled	Time-scaled	Time frame of EMP
Contractors	Require compliance		Communication only.
Documentation	Register of effects		Not specific
Audits	Not specific	3 times a year	Not specific
IER		Required	Suggested
Communication	Only Policy statement	Public statement	Only policy

Source: SDPI Database, 2001.

8

Environmental Standards and Exports of Bangladesh

ZAID BAKHT

SECTION I

Introduction

The world economic order has undergone significant changes over the past decade or two. Rapid developments in information and communication technology greatly reduced the cost of transactions across national frontiers. The result has been sharp increase in international flow of goods and service. At the same time, the successful completion of the Uruguay Round of GATT negotiations and the emergence of WTO as the successor of GATT provided strong impetus to move towards the multilateral rule-based global trade regime.

However, this move towards global integration has been marked by great unevenness. The advanced market economies and the Newly Industrialised Countries (NICs) of East Asia account for a disproportionate share of the enhanced trade flows relative to their share in global income. Between 1980 and 1990, the share of the Least Developed Countries (LDCs) in the global trade declined from 0.8 per cent to 0.4 per cent and remained stagnant at that level throughout the decade of the 1990s. The experience bears testimony to the fact that for a developing economy a positive outcome originating from the process is neither automatic nor guaranteed.

The standard prescription for an economy striving to integrate with the global economy is liberalisation of the internal and the external market to facilitate supply-side flexibility and attain macro-economic stability. Apart from trade policy reforms, this would involve openness to capital and technology flows, unhindered flow of domestic investments and labour across sectors, financial reforms to permit market determination of investment and saving, and public sector disinvestment. These reforms seek to reallocate resources from the production of non-tradables to tradables. Improved resource efficiency is secured from the exposure of enterprises to internal and external competition and through a drastic reduction in the scale and discretionary component of government interventions in enterprises and markets. In short, the thrust of this prescription is on 'getting the price right' through unfettered market.

However, globalisation of the world economy is changing the dynamics of world trade and this in turn is changing the focus of competitiveness from price aspects to structural aspects. What counts more in the current competition between nations is the quality of the goods and services produced and the skill of the human resource involved in such production. The pattern of world trade is increasingly being dictated not by comparative advantage based on factor endowments and cheap labour but by competitive advantage determined by quality, speed of delivery and presentation. Essential ingredients of competitive advantage are technological innovation, training and systematic updating of human resources, management techniques compatible with the global economy, flexibility to adapt to the ever changing global marketplace, and creativity. The structural aspects of competitiveness again depends on a whole set of systems involving suppliers of goods and services, the financial, educational, technological, energy, transport, and telecommunications systems, infrastructure and the quality of the public institutions. World competition today is not between individual firms but between these systems. Shortcomings in these areas all affect the competitiveness of the firm. It is not surprising, therefore, that developing countries beset with important structural problems such as low skill capacity, poor infrastructure, weak institutions, low levels of investment and a weak state of governance are finding it increasingly difficult to cope with the challenges of globalisation.

The woes of the developing countries are further compounded by threats of market restrictions based on social concerns. One such non-trade issue looming high is environmental standards. The demands from many high-income economies to link such social issue to trade policy are based on the following arguments. First, they are concerned that lower environmental standards provide industries in developing countries with a competitive advantage over their own industries. Second, there is the concern of international spillover effects resulting from environmental degradation. Thus, trade policy is seen by some as means of addressing these competitive and social concerns.

The linking of trade policy with environmental policy is likely to create several difficulties for the developing economies. First, there remains the risk of such social clauses being used for protection against competitive imports from developing world. Second, the capacity of the developing countries to adopt better environmental standards are often restricted by structural and resource constraints. The imposition of more stringent environmental condition, therefore, is unlikely to result in improvement of such standards and instead would marginalise these countries in world trade by undermining their comparative advantage. To the extent that environmental practices are often function of the level of development, the better way to address environmental objectives would perhaps be to generate economic prosperity through greater market access and trade.

This chapter attempts to examine the possible impact of enhanced environmental standards on the external trade of Bangladesh. As the preceding paragraphs have suggested, these impacts are likely to depend on (a) the underlying strength of the export sector, (b) success in eliminating policy induced constraints, (c) the structural bottlenecks to export growth, and (d) environment sensitivity of the export industries and the ability to cope with these.

Accordingly the chapter is organised as under. After the introductory remarks in Section I, Section II provides an overview of Bangladesh's external sector highlighting its strength and weaknesses. Section III gives a brief account of the policy reforms carried out by Bangladesh in liberalising its markets. The environmental impacts of exports are discussed in Section IV while Section V reviews the incidence of relocation of industries in Bangladesh due to less stringent environmental standards. Finally, concluding remarks are presented in Section VI.

SECTION II

The External Sector of Bangladesh

Growth Performance

Between 1980-81 and 1990-91, Bangladesh's total exports in current US $ value grew at an annual compound rate of 9.2 per cent (Table 8.1). The annual growth rate was 14.4 per cent during the period 1990-91 to 1999-00. In fact, Bangladesh experienced double digit export growth in most of the years during the 1990s. However, the growth rate took a nosedive to 2.9 per cent in 1998-99 due to adverse impact of prolonged flood. Export growth revived to 8.3 per cent during 1999-2000.

Table 8.1

Exports and Imports of Bangladesh during 1980s and 1990s

Description	*1980-81*	*1990-91*	*1999-00*	*Annual Compound Growth Rate (%)*	
				1980s	*1990s*
Export (Million $)	710	1718	5752	9.2	14.4
Import (Million $)	2282	3510	8566	4.4	10.4
Trade Deficit (Million $)	1572	1792	2814	1.3	5.1
Export as % of Import	31.1	48.9	67.1		
Export as % of GDP	5.0	7.3	12.1		
Import as % of GDP	16.0	15.0	17.9		
Openness of the Economy (%)	21.0	22.3	30.0		

Source: Export Promotion Bureau and World Bank.

Imports, on the other hand, grew at an annual compound rate of 4.4 per cent and 10.9 per cent during 1980-81 to 1990-91 and 1990-91 to 1998-99 respectively. The gap between export and import widened from $1572 million in 1980-81 to $1792 million in 1990-91 and further to $2814 million in 1999-00, although the share of export earnings in import payments steadily rose from 31 per cent in 1980-81 to 67 per cent in 1999-00.

The openness of the economy as measured by total external trade as a proportion of GDP went up from around 21 per cent in 1980-81 to nearly 30 per cent in 1998-99 with the share of export in GDP rising from 5 per cent to 12 per cent during the same period.

Changing Structure of Export

The structure of export has changed significantly over the past two decades. Bangladesh seems to have made the transition from resource-based to process-based exports. In 1980-81, primary commodity constituted nearly 29 per cent of total exports. In 1990-91, this share came down 17.8 per cent and further down to 7.9 per cent in 1998-99 (Table 8.2).

Table 8.2

Changing Structure of Export (1980-81 to 1999-00)

(Million $)

Item	*1980-81*	*1990-91*	*1999-00*	*Growth (%) 1990-1999*
Primary Commodities	209 (29.4)	306 (17.8)	469 (8.2)	4.9
Raw Jute	119 (16.8)	104 (6.1)	72 (1.3)	Negative
Tea	41 (5.8)	43 (2.5)	18 (0.3)	Negative
Frozen Food	40 (5.6)	142 (8.3)	344 (6.0)	10.3
Other Primary	9 (1.3)	17 (1.0)	35 (0.6)	8.4
Manufactured Goods	501 (70.6)	1411 (82.2)	5283 (91.8)	15.8
Jute Goods	367 (51.7)	290 (16.9)	266 (4.6)	Negative
Leather & Leather Goods	57 (8.0)	136 (7.9)	195 (3.4)	4.1
Woven Garments	3 (0.4)	736 (42.9)	3083 (53.6)	17.3
Knitwear	0 (0.0)	131 (7.6)	1270 (22.1)	28.7
Chemical Products	11 (1.5)	40 (2.3)	94 (1.6)	9.9
Other Manufactured Goods	63 (8.9)	78 (4.5)	375 (6.5)	19.1
Total Export	710 (100)	1717 (100)	5752 (100)	14.4

Note: Figures within parentheses show percentage share in exports.

Source: Export Promotion Bureau.

However, when one takes a closer look at the disaggregate picture, the feature that strikes the eye is the shift from jute-centric to garments-centric export. In 1980-81, raw jute and jute goods together constituted 68 per cent of total exports. Between 1980-81 and 1998-99, export of both raw jute and jute products declined in absolute terms and their total share came down to only seven per cent in 1998-99. In contrast, woven and knit garments together accounted for less than one per cent of exports in 1980-81. Their combined share in exports rose to nearly 76 per cent in 1998-99.

The evidence presented brings out the basic weakness of Bangladesh's export sector. In spite of the rapid growth in exports, particularly of manufactured exports, the export base has remained extremely narrow. In fact, it has become narrower during the 1990s. The observed growth in export has been propelled mainly by growth in the export of garments—both woven and knit. Two other items that registered significant growth during the 1990s are frozen food and chemical products.

Market Concentration of Exports

The commodity concentration also led to concentration in the destination of exports. Historically, the major destinations of Bangladesh's export have been USA, EU, Canada, Japan and China. In 1980-81, these regions received about 30.5 per cent of Bangladesh's export. But market concentration accentuated over time and in 1998-99, the same region was the destination for 88.0 per cent of Bangladesh's export (Table 8.3).

Table 8.3

Percentage Distribution of Exports by Destination

Destination	*1980-81*	*1990-91*	*1999-00*
USA	12.3	29.5	39.5
EU	10.6	39.1	42.7
Canada	0.9	1.8	1.9
North America & EU	23.8	70.4	84.1
Japan	2.7	2.4	1.7
China	4.0	1.9	0.2
Sub-total	30.5	74.7	86.0
ASEAN & NIEs	5.5	4.3	3.8

Source: Export Promotion Bureau, 2001.

If one looks behind the export performance of Bangladesh, one finds that the spectacular growth of the garments industry was triggered by the relocation of their trade in Bangladesh by East Asian exporters in an attempt to circumvent binding quota restrictions in their own country imposed under the Multi-Fiber Arrangement (MFA) particularly for the North American market. This resulted in the direction of Bangladesh's garments export to remain focussed towards USA and Canada.

The other factor that contributed towards the rapid growth of apparel sector particularly of knitwear in Bangladesh was the zero-tariff access to the EU markets under the GSP. Because of these twin benefits of MFA and GSP, 97 per cent of all apparels exports from Bangladesh were directed towards North America and EU markets even in 1998-99.

Two issues are of concern in connection with direction of export: (a) the spread of the export market, and (b) the share of growing economies in the export market. If the export of a region is narrowly directed to a handful of economies, it remains vulnerable to fluctuations in the economic performance of these economies. Again, a larger share of the developed and growing economies in the export market indicates a higher incidence of income-elastic products in the export basket and hence higher potentials for growth of export along with the growth of these economies.

Judged on those criteria, Bangladesh seems to have failed to broaden her export market. She was particularly unsuccessful in penetrating the market in the growing economies of ASEAN and the NIEs. The share of her export going to this region declined from 5.5 per cent in 1980-81 to 3.8 per cent in 1998-99.

Trend in Export and Import Price

As would be expected, Bangladesh's exports have failed to climb up the value ladder significantly during the past two decades. The readymade garments industry, which has been the prime mover of exports during this period, remained concentrated at the relatively lower end of the market segment producing mainly T-shirts, pajamas, ordinary shirts, shorts, caps, women's and children's wear, etc. Diversification into higher valued jackets, shirts of complicated designs, branded items, etc. has been slow. Similar has been the experience with the export of leather goods. Jute goods, another important item of manufactured exports, lost in unit value due to displacement by synthetics and slowing down of world demand. In the case of primary commodities such as raw jute, tea, and frozen food, export price index virtually stagnated. The overall export price index, thus, registered only modest rise during the past two decades (Table 8.4). With the import price index rising by a higher margin, the terms of trade deteriorated during the 1990s. The situation became particularly bad in 1998-99 when all major export items of Bangladesh except woven RMG experienced decline in unit value by a margin of 5-40 per cent. As a result the overall export price index rose marginally by 0.6 per cent. This was largely the aftermath of East Asian financial crisis that led to worldwide fall in commodity prices, and it contributed significantly to the sharp fall in Bangladesh's export earnings in that year.

Table 8.4

Price Index of Export and Import

Description	*1980-81*	*1990-91*	*1999-00*	*Yearly Growth (%) 1979-1999*
Export Price Index	100	118.55	140.6	1.7
Import Price Index	100	111.25	158.0	2.3
Terms of Trade	100	106.56	89.0	Negative

SECTION III

Bangladesh carried out wide-ranging trade policy reforms during the past two decades. The reforms covered both tariff and non-tariff barriers. With regard to non-tariff barriers, the focus has been on deregulation of import procedure and elimination of quantitative restrictions. In the case of tariff barriers, the attempt has been to rationalise the tariff structure, reduce the number of duty slabs, and bring down tariff rates and their dispersion amongst similar commodities.

Non-tariff Barriers

Bangladesh has come a long way in terms of dismantling its non-tariff barriers. The system of import licence has been virtually eliminated and the Control List of Banned and Restricted items has been brought down to a minimum. As part of the conditionality of concessionary loan under the World Bank Industrial Sector Adjustment Credit (ISAC-I), it was stipulated that beginning in 1987-88, there would be a yearly reduction of 20 per cent in the Negative list. Table 8.5 shows the change in the level of quantitative restrictions between 1986-87 and 1997-2002 measured at the four-digit level. The main manufacturing sector receiving protection from QR on trade-related ground is the textile sector, with a quarter of tariff lines in this sector still remaining under QRs.

Intervention by state trading enterprises has been reduced over time. The role of the state trading body—the Trading Corporation of Bangladesh (TCB)—in import and export has been curtailed, and there is now wider participation by the private sector in import and export trade that were earlier reserved for the TCB.

Table 8.5

Number of 4-digit H.S.-Lines under Quantitative Restrictions

Item	*1986-87*	*1990-91*	*1997-2002*
Total	550	239	124
Trade Reasons	489	179	28
a. Banned	252	93	5
b. Restricted	151	47	6
c. Mixed	86	39	17
Non-trade Reasons	61	60	96

SECTION IV

Impact of Environmental Standards on Exports

Bangladesh is a low-industrialised country as reflected in the low share of the manufacturing sector (15 per cent) in the GDP and low growth of manufacturing value added (4 per cent) attained over the past three decades. But the environmental degradation caused by the sector is quite high. The pollution absorption capacity of the country is also quite low given that the country has a small land area but a large population. Industries that are major contributors to manufacturing GDP and exports are either natural resource based processes or pollute the environment through emission of waste and toxic products. Water pollution through untreated discharge of solid wastes and liquid effluents and air pollution through gaseous emissions are the most pervasive forms of industrial pollution in Bangladesh. According to the Department of Environment (DoE), the major polluting units belong to the following industrial sub-sectors:

i) Shrimp culture,

ii) Sugar,

iii) Distilleries,

iv) Textiles including dyeing & finishing,

v) Leather tanning,

vi) Pulp and paper,

vii) Fertiliser, and

viii) Re-rolling mills.

Amongst these polluting industries, shrimp culture is a fully export oriented industry while leather tanning and fertiliser cater to both domestic and export markets. Textiles including dyeing and finishing are important backward linkage industries for Bangladesh's main export earner readymade garment industry. The following sub-sections provide a brief account of these export industries and the nature and magnitude of environmental degradation caused by them.

Shrimp Culture

Profile of the Industry

In 1980-81, export of frozen food from Bangladesh was only US $ 40 million. In 1999-2000, it reached the level of US $ 344 million, the annual compound growth rate being nearly 12 per cent. Although the share of this item in total export is currently about 6 per cent, in value added terms it is the single largest export of Bangladesh based almost entirely on local raw materials. Nearly 89 per cent of the export of frozen food from Bangladesh consist of shrimp.

Up to mid-1980s, the export was based mainly on open water catches of shrimp. Coastal aqua-culture for export of shrimp is predominantly a development of the late 1980s and early 1990s. Rising world prices provided the incentive to venture into commercial shrimp farming. Price of shrimp went up from $2.4 per pound in the mid-1980s to $4.5 per pound in the mid-1990s. Given that the product is largely natural resource and labour based, Bangladesh had a comparative advantage in the export of shrimp. A World Bank/ UNDP investment programme of $30 million also helped launch the export oriented commercial shrimp farming with infrastructure and technology during this period. The government also joined in with a number of incentives covering amendments to the land-lease laws, subsidised credit and tax holidays. A significant portion of the shrimp export now originates from the aqua-culture sector. Like other exports of Bangladesh, the export market for shrimp is also narrow with about 55 per cent going to USA followed by EU (19 per cent) and Japan (14 per cent).

Environmental Consequence of Shrimp Culture

Shrimp culture involves prolonged inundation of land by saline water. This causes salinity of the land to increase and soil fertility to decline not only in the land under shrimp culture but also in the neighbouring lands. As a result, production of agricultural crops and fodder decline sharply. The salinity level in neighbouring fresh water ponds also rises creating scarcity of fresh water for drinking and household purpose. It also contributes to public health hazards such as deposition of salt on human skin and higher incidence of water-borne diseases.

Shrimp fry used for shrimp culture are collected from natural sources, namely, coastal rivers, seacoast and estuarine area. During the harvesting of each shrimp fry, about 90-100 eggs or fries of other aquatic species get destroyed, which affects the ecological and biological system in the coastal area. Moreover, during transport of fries from harvesting area to shrimp culture ponds, a large percentage of the fries are lost unless handled properly.

Ecological systems are also affected by harvesting of shrimp by trawlers unless care is taken to extrude other species. There are also complaints of over-trawling and harvesting in breeding areas prohibited by law.

Product Quality and Health Standards

Internationally recognised product and health standards such as ISO certification, HACCP, etc. are being increasingly demanded of the exports of the developing countries. Current state of product quality monitoring and control in these countries falls far short of international standards. As a result, enhanced health and hygiene are having similar trade restrictive effects on the exports of these countries as enhanced environmental standards. In fact, the distinction between the two seems to increasingly getting obscure. The case of EU's ban on import of shrimp from Bangladesh for non-compliance of HACCP regulations highlights the urgent need for technical assistance to support the export industries in upgrading their product quality and hygiene standards and for improving the institutional capability to monitor and control such standards.

European Commission Ban

In July 1997, the European Commission imposed a ban on import of frozen shrimp from Bangladesh on the grounds of non-compliance with the importer's hygienic regulation HACCP.[1] Specifically, the objections related to the following:

(a) use of unskilled workers with unhygienic habits,

(b) unhygienic transportation and preservation,

(c) untimely procurement of shrimp, and

(d) corrupt practices for making excessive profit.

The EC had initially given Bangladesh up to November 31, 1997 for taking adequate measures to ensure compliance with the regulations. On the basis of a visit by an EC technical team the time limit was extended by another three months. The ban continued for nearly eight months. In February 1998, the EU lifted the ban on imports from six shrimp processing firms, and from another 11 firms in July 1998. Thereafter, the EU ban was gradually lifted from a large number of other exporters as those plants were eventually upgraded. A total of $17.6 million was spent in upgrading the plants and the industry cost required to maintain HACCP is estimated to be $2.2 million per year.

1. HACCP is a system of regulations that has been widely accepted as an international standard to control all factors related to contamination, survival and growth of microorganisms in food in all stages of the food chain. It is based on the recognition that microbiological hazards exist at various points of the food production and distribution chain, but measures can be taken to control these hazards at specific points in the chain.

Given that, EU is the single largest buyer of Bangladeshi shrimp, the ban spelled disaster for the industry. Nearly 80 per cent of the shrimp processing plants got shut down. The following table show how the export of frozen shrimp declined steeply after the imposition of the ban.

Table 8.6

Export of Frozen Shrimp from Bangladesh to the EU

Month	*Export ($Million)*
August '97	46.4
September '97	38.8
October '97	30.9
November '97	25.4
December '97	19.6
January '98	11.9
February '98	8.1

Impropriety of the European Ban

While the importance of adequate health and hygienic standards in the export of food items can hardly be denied, the manner in which the EU ban was imposed on the import of frozen shrimp from Bangladesh reflects its arbitrary nature and also the vulnerability of the weak economies such as Bangladesh to non-tariff barriers against their exports.

It is generally known that there are about 25-30 firms in Bangladesh who are well equipped and who export quality shrimp under proper hygienic conditions. The EU has all along been importing from these firms as well as from other firms in Bangladesh. Thus, EU was aware of the variations in the quality of the shrimp processing units in Bangladesh. The government of Bangladesh was informed of the problem of poor hygienic quality of shrimp processing. An FAO project aimed at assisting the formulation and implementation of a HACCP based assurance programme was already underway. The EU imposed the ban after it visited a few firms recommended by the government and found them to be of unsatisfactory hygienic standards. Ideally, the EU should have continued importing from the quality exporters instead of imposing the ban across the board. They should have provided appropriate guidelines for the improvements of the other firms and given them lead time to improve. EU could also provide technical assistance where necessary. Only in the event of failure of these firms to comply with the health and hygiene requirements could EU impose ban on import from these firms.

The above experience epitomises the difficulties Bangladesh faces in coping with the provisions of the UR Agreement on the Applications of the Sanitary and Phyto-sanitary measures. The agreement seeks to introduce harmonised standards in respect of health and safety regulations for the protection of human, animal and plant life or health. The basis for harmonisation adopted in the agreement are three international standards. For food safety, the standards, guidelines and recommendations established by the Codex Alimentarius Commission, relating to food additives, veterinary drugs and pesticide residues, contaminants, methods of analysis and sampling, and codes and guidelines of hygienic practice are to be followed. While the animal health standards are to follow the guidelines and recommendations of the International Office of Epizootics, plant health standards are to set in line with the international standards, guidelines and recommendations developed by the Secretariat of the International Plant Protection Convention (IPPC) in cooperation with regional organisations operating within the framework of the IPPC.

Scope of Containing the Environmental Fallout of Shrimp Culture

The extent to which agricultural production is displaced by shrimp culture will depend on the level of salinity and soil fertility that existed prior to the setting up of the shrimp farm. A zoning policy that demarcates the area in terms of high, medium and low saline zones can, therefore, be helpful in containing the displacement effect of shrimp culture.

The extent of salinity and the level of degradation of soil fertility have been found to vary significantly with the type of shrimp culture practiced. The environmental fallout of shrimp culture has been found to be much less when intensive or semi-intensive types are used and are integrated with mangrove/paddy plantation, etc. as may be appropriate. What is also important is that the productivity of intensive shrimp farms is much higher than the productivity of the extensive type farms. Unfortunately, there has been indiscriminate horizontal expansion of extensive shrimp culture in Bangladesh, which has not only caused significant environmental damage but has also resulted in the productivity level to be less than one tenth of what could be achieved even with semi-intensive practice.

The damage to ecosystem due to procurement of fries from natural sources can be easily corrected through establishment of modern hatcheries. In the meantime, training of fry collectors through adequate extension service can help minimise the spoilage of both shrimp fries and fries/eggs of other species. Similarly, proper implementation of trawling rules can help contain damage to sea-resources.

A review of the present health inspection and quarantine rules of Bangladesh shows that they are totally out-of-date and needs to be drastically modified for compliance with the UR Agreement. In the specific context of shrimp export, the Department of Fisheries needs to take several steps to meet the HAACP regulations. These include: (i) Building hygienic service centres near the shrimp farms, which are needed to prepare shrimp for transportation to the processing plants by washing and sorting it, and by packing it into baskets with ice. (ii) Modifying design and operation of the processing plants to meet the regulations. (iii) Importing additional refrigerated trucks. (iv) Continuing the training workshops and technical assistance programmes already undertaken by FAO.

Recognising the limitations of the developing countries in meeting the requirements of regulations, the agreement on the Applications of the Sanitary and Phyto-sanitary measures stipulated that they should be granted, upon request, specified time-limited exceptions, in whole or in part, from obligations under the agreement, keeping in view their financial and developmental needs. The unilateral action of the EU in banning shrimp import from Bangladesh goes against this spirit of accommodation of the developing countries concern.

Leather Tanning

Profile of the Industry

Bangladesh produces about 182 million square feet of hides and skin every year of which only about 15-20 per cent is used up domestically. Given the availability of the basic raw material for leather production and the labour-intensive nature of the activity, leather and leather goods manufacturing constitute important export-oriented industries for Bangladesh. In 1980-81, leather export from Bangladesh was to the tune of US $ 57 million only and consisted almost entirely of semi-processed leather called wet blue. To achieve greater value addition, the government set July 1990 as the limit beyond which export of wet blue leather would be banned and provided fiscal and financial incentive for up gradation of the leather units for production and export of more processed crust and finished leather and leather products. In 1991-92, leather export amounted to US $ 144 million and all of it was in the form of crust or finished leather. At around that time, a few firms ventured to produce selected leather goods for overseas market and exported leather goods worth US $ 4 million in that year. During the mid-1990s, a few leather footwear manufacturing units with modern machinery were set up and started production of quality shoe for selected overseas markets. In 1997-98, export of leather goods including footwear stood at US $ 48 million registering an annual compound growth rate of nearly 51 per cent. In 1999-00, total export of leather sector amounted to $195 million (3.4 per cent of total export) of which 65 per cent were crust leather, 27 per cent finished leather and 8 per cent leather products. Leather export also has a concentrated market with 35 per cent going to Hong Kong, 24 per cent to Italy, 6 per cent to Japan and another 4.9 per cent to Spain.

Environment Sensitivity of the Leather Sector

Leathery tanning industry in Bangladesh faces serious pollution problem on the domestic front. Of the 214 tanneries in the country, 196 units are located over 25 hectare of land in an area called Hazaribag in Dhaka city and they process nearly 95 per cent of all hides and skins. All these units drain toxic wastewater in the river Buriganga. In addition to the wastewater, the tanneries also discharge large volume of highly toxic effluence into the river. The most hazardous element in the waste disposal of the tanneries is chromium, which in its hexavalent form is carcinogenic. Concentration of chromium in Buriganga River is observed to be 6mg/l whereas the maximum allowable limit is 4 mg/l. The industry also causes air pollution while soaking, processing and finishing leather through the discharge of harmful chemicals and strong odour.

On the external front, restrictions on the use of certain chemical dyestuffs particularly PCP causes difficulties for the industry. EC standard for PCP is 10 mg/kg, which is followed strictly by Germany but less so by Italy, Spain and France. The restrictions imposed by Hong Kong are also less stringent. Since Germany accounts for less than two per cent of Bangladesh's leather export, problems faced by Bangladeshi exporters has not been very acute so far.

Containing Environmental Hazards of the Leather Industry

Leather industry in Bangladesh is dominated by small and medium enterprises. Given their weak financial standing, these units have special problems in complying with higher environmental standards. The small and medium enterprises have to depend on the support from the government and other agencies to address the environmental problems of leather industry. There are two important steps that are required for containing the environmental hazards of the leather industry. First, the provision of effluent treatment plant to free the waste from toxic substances and second relocation of the industry at a suitable place away from the city. Currently, only a handful of tanneries are known to treat its wastewater effluent in an effective manner. The government had taken up a plan to set up a leather estate in the outskirts of the city. But because of the financial cost involved and possible dislocation in exports, the entrepreneurs have been less enthusiastic about the leather estate, and not much progress has been made so far in this regard. The government has recently undertaken a programme to establish a common effluent treatment plant (CETP) in Hazaribagh with the assistance of UNIDO. Roads and drainage systems in the tannery area are now being improved for proper operation of the CETP.

Fertiliser Industry

Profile of the Industry

Yearly production of chemical fertiliser in Bangladesh is approximately two million mt. About 75 per cent of this fertiliser is used domestically and about half a million tonne is exported annually. Export earnings from fertiliser was US $ 10.3 million in 1981-82 and increased to US $ 104.1 million in 1996-97 but came down to US $ 59.9 million in 1999-2000. Yearly compound growth of fertiliser export in current dollar value between 1981-82 and 1999-00 works out at 10.3 per cent. However, in terms of proportion of total export earnings, the share of fertiliser remains miniscule at 0.1 per cent. Major export destinations of fertiliser are Australia (26.2 per cent), India (13.5 per cent) and Thailand (6.2 per cent).

Nature and Magnitude of Pollution Caused by the Fertiliser Industry

Except one joint venture, all the fertiliser plants in Bangladesh are in the public sector are set up with foreign aid. The foreign aid involved tied technical assistance and equipment acquisition from the donor country. The environmental performance of these units, therefore, reflects donor-country standards at the time of installation. As would be expected, the more recent plants are cleaner than the earlier ones as they embody more recent technology consistent with stricter environmental standards.

The oldest urea fertiliser plant (Natural Gas Fertiliser Factory), built with Japanese assistance in 1961, is located in the Northeastern part of the country on the Kushiara River. The plant's discharges contain toxic ammonia, high BOD and PH levels, and substantial residues of grease and oil. They have been clearly identified as the cause of major fish kills, paddy field damage and health threats. In addition, there have been regular atmospheric releases of ammonia and sulphur di- and trioxide. The intensity of pollution by this plant is ranked as high amongst all the urea fertiliser units in the country. A second (UFF) and third (PUFF) urea plant set up in 1968 and in 1985 with Japanese and Chinese assistance respectively have comparable technology. Both plants are clearly identifiable polluters the pollutant being same as the earlier plant, namely, BOD, Ph, ammonia, urea, alum sludge, oil and grease. However, in terms of intensity of pollution these two plants fall in the middle range. In spite of being of a later vintage, a Triple Super Phosphate Complex (TSPC) set up in 1970 in Chittagong is identified as the most pollution-intensive dangerous fertiliser plant in the country. The wastewater is loaded with flouride and sulphur. There are heavy emissions of sulphur di- and trioxide, flouride and nitrous oxide. The Chittagong Urea Factory set up in 1989 with Japanese assistance marked the beginning of high standard of process cleanliness. The plant internalises environmental concerns to such an extent that effluent load of wastewater is very low and there is little need for 'end-of-pipe' waste treatment.

Containing Pollution Impact of the Fertiliser Industry

It is recognised that the age and technology of NGFF preclude clean-up to a very high standard. The government has, therefore, settled for a modest first-level clean up effort. Earlier, effluents went directly into the Kushiara River. Now, there are two lagoons in which the effluent is diluted by adding the staff colony's wastewater and local spring water. In addition, acidic and alkaline injections are used to neutralise the PH level. Some separation of oil and grease from the water has also been undertaken. Community pressure has also resulted in payment of some compensation and prior warnings before periodic cleaning operations discharge large quantities of ammonia and other pollutants in the river.

In the case of UFF and PUFF, there were strong community reaction to pollution and the qualitative response of the two plants were similar to NGFF. UFF has increased the number of employees working on pollution control. Both pay some local compensation for damage claims. Both plants share a first stage treatment lagoon to dilute the effluent with wastewater from their staff colonies. UFF employs urea hydrolysis, an ion exchange facility and an oil/grease separation plant. PUFF reduces the ammonia load in its effluent with a steam stripping method, and spreads a simple cloth barrier over the out-fall to capture some of the oil and grease. In the case of TSPC, however, little clean up efforts were undertaken in spite of strong community protests. The main reason being that the technical and human resource requirements for effective clean up in the facility was much more demanding. In part, it was also a management failure.

Textile Industry and Readymade Garments

Textiles is by far the most dominant manufacturing activity in Bangladesh. There is a pervasive presence of textiles across all size classes—large, medium, small and cottage industries. Textile industries have also been identified as the third major polluting industries in Bangladesh. Major pollutants in textile industries are wastewater containing alkali, chlorine dye, organic matters, detergents, etc. Pollutants are particularly high in dyeing and printing activities. Pollution load of a typical composite textile mill in Bangladesh is BOD 500mg/l, COD 500 mg/l and total suspended solid 375 mg/l, while Bangladesh standards for such pollutants are 50 mg/l, 200 mg/l and 150 mg/l respectively.

Bulk of the textile industry of Bangladesh is, however, domestic market oriented and hence stringent textile related environmental conditions of developed countries are unlikely to have any significant impact on the polluting effects of Bangladeshi textile industries. Currently, less than 15 per cent of the fabric used in the export oriented readymade garments industry are met from domestic sources.

The readymade garments industry, however, faces strict regulation in the European market regarding dyes and chemicals used in the fabric. Germany particularly prohibits apparels containing azo-dyes. Given that, EU in general and Germany in particular is a major importer of Bangladesh's readymade garments, the RMG exporters are obliged to comply with these requirement. Since Bangladesh imports bulk of dyes, chemicals and textiles that are used to feed its RMG industry, the stringent regulations of EU does not put her into particular cost disadvantage. Discussions with RMG exporters revealed that they see the myriad of environmental standards and regulations more as business irritant. Since testing facilities are not available domestically, the exporters have to procure from abroad certificates regarding absence of particular types of dyes and chemicals sometimes incurring certain amount of cost.

SECTION V

Migration of Dirty Industry to Bangladesh

One anticipated adverse effect of enhanced environmental standards is the possible relocation of polluting industries in countries where environmental measures are lower or weak. To what extent has this happened in Bangladesh?

It was noted in the earlier section that Bangladesh's export of shrimp grew at the fairly rapid rate ten per cent per anum during the 1990s. The comparative advantage of Bangladesh in shrimp production is revealed by the fact that the share of Bangladesh in world export of shrimp increased from around two per cent in 1990 to nearly four per cent during late 1990s. This means that Bangladesh's export of shrimp grew at a higher rate than world export during this period. However, a major factor contributing to rapid growth of shrimp production in Bangladesh was the implementation of a World Bank/ UNDP investment programme of $30 million, which helped launch Bangladesh's shrimp export industry with infrastructure, technology and foreign advice. The government also came forward with a number of proactive policies including amendments to the land-lease laws, subsidised credits and tax holiday. Thus, the rapid expansion of shrimp industry in Bangladesh during the 1990s was largely due to endogenous reasons and there is no evidence to suggest that weaker environmental standards of Bangladesh contributed to this process in a major way.

The leather sector also does not seem to have been influenced by relocation of industries. Bangladesh's export of leather and leather goods increased at a modest annual rate of 4.1 per cent during 1990s. Compared to tannery, leather goods manufacturing is less dirty as the process and production method are less polluting. In 1990, leather goods constituted only two per cent of total export earnings from leather and leather goods. In 1999-00, this share rose to eight per cent. Thus, expansion of the more dirty component of the leather sector has been rather low in Bangladesh during 1990s and hence deficient environmental standards does not seem to have resulted in relocation of leather industries in Bangladesh during this period.

Bangladesh's export of fertiliser increased at an annual compound rate of nearly 10 per cent during 1990s against the growth rate of 5 per cent in world export of fertiliser. However, since 1990-91, there has not been any new investment in the fertiliser industry in Bangladesh and the annual production of urea fertiliser oscillated between 1.5 and 2 million metric tonnes during this period. Thus, no relocation of fertiliser industry seems to have been prompted in Bangladesh during this period due to differences in environmental standards.

The spectacular growth of export oriented readymade garments and knitwear industries in Bangladesh during the 1980s and the 1990s was triggered by the relocation of the trade by international intermediaries faced with binding quota restrictions in their own country. At the same time, quick government support in the form of duty free imports, back-to-back LC facility, XPB benefit, duty relief on import of machinery, etc. helped significantly to create the needed incentive. The continuation of the protection provided by the country specific import quota, specially in the USA, under the MFA and the facilities provided under the GSP gave the main stimulus to the growth of this industry. Thus, the rapid growth of readymade garments and the various backward linkage industries including textiles in Bangladesh seems to have been the result of some special factors relating to world trade in these commodities rather than the impact of varying environmental standards.

SECTION VI

Concluding Remarks

The evidence presented in this chapter has shown that the environmental hazards of Bangladesh's exports are quite high. The review of four major export items, namely shrimp, leather, fertiliser and RMG, however, suggest that cost considerations are not the main reasons behind the current environment degrading features of these activities. Rather, lack of awareness and technical know-how, infrastructural deficiencies, management failures, inadequate enforcement of environment related laws, absence of supportive role of the State particularly in providing extension service, common effluent treatment plant, etc. are mainly responsible for the persistence of the environment related ills of these industries. Therefore, mere application of stringent international environmental standards is unlikely to be successful in bringing about an improvement in the basic malice. On the contrary, the evidence suggests that in spite of the impressive growth performance, the external sector of Bangladesh has remained extremely narrow and un-diversified both in terms of commodity composition and market orientation. Bangladesh, therefore, remains highly vulnerable to unilateral imposition of trade restrictions on such non-trade grounds. Instead of punitive actions, there is need for technical and financial support from the developed importing countries for Bangladesh to overcome the structural constraints to export diversification in general and solving the environmental ills of her export industries in particular.

9

Case of Sri Lanka

SAMPATH JAYASINGHE

Introduction

When the international trading system was reconstructed after the Second World War, the environmental consequences of economic integration was not a primary concern. Only indirect references to the environment were included in the exception clause of GATT 1947, Article XX, which allows countries to sidestep the normal trading rules if necessary to protect human, animal or plant life or health, or to conserve exhaustible natural resources, provided that such measures do not discriminate between sources of imports or constitute a disguised restriction on international trade. After a long silence the issue was put back on the agenda first in the early 1970s in the preparation for the UN Conference on Human Environment, held in Stockholm in 1972. However, no significant commitment was made to mitigate the issues on trade and environment until the beginning of the 1990s.

One reason for the low environmental profile of the GATT in the formative decades was that trade was not perceived to be an environmental issue as such, neither among policymakers nor the public at large. What perceived was at that time according to theoretical background, if governments used proper environmental policies to internalise the environmental cost of production and consumption, trade would eventually raise welfare (McGuire, 1982). But environmental policies lagged behind in many countries. Therefore, the basic presumption on which the free trade case rested was not satisfied. This background was not seen as a reason to halt or reverse the process of trade liberalisation, but to rather strengthen environmental policies and institutions, including at the international level.

As environmental issues had a low priority during the first four-decade, of the GATT, they came back with a vengeance in the early 1990s. The starting point of the current debate was a series of continuous environment related trade disputes, especially the 'tuna dolphin' dispute between Mexico and the United States. With the formation of the WTO in 1995, environmental issues, as they related to trade, are now firmly anchored in the multilateral trading system. The objective of WTO, as spelt out in the preamble of the Marrakesh Agreement establishing the WTO, now embraces the internationally recognised principle of sustainable development, defined by the World Commission on Environment and Development in 1987 (Nordstrom and Vaughan, 1999).

The environment community argues that free trade is responsible for some of the environmental problems as free trade means more production and consumption of environmentally unfriendly products, which result in more pollution and environmental degradation. Trade policy instrument such as tariff and export taxes have been suggested to minimise the damage caused by such trade. Supporters of free trade argue that free trade generates gains from specialisation and exchange to the countries involved in such trade and that additional income received can be used for more environmental protection. Pro-trade groups argue that the environmentalists' demands undermine the prospects of mutually beneficial trade.

The concerns for environmental protection are widespread in the countries such as Sri Lanka where rapid trade liberalisation programmes are in place. Since 1977, Sri Lanka has promoted export-led growth as a means to achieve rapid economic development. Although Sri Lanka has made tremendous progress in trade liberalisation, along with many other developed and developing countries, no attempt has been made so far in policymaking circles either to examine the implications of trade liberalisation on environment or to reconcile trade and environment policies. The purpose of the chapter is to examine some select issues involved in trade and the environment giving more emphasis on 'how environment impacts on trade'. In attempting to investigate the link between trade and the environment, the chapter is designed to examine firstly, the relationship between environmental management and trade drawing some evidence from literature. Secondly, it reviews effects of complying with domestic environmental standards and requirements on Sri Lanka's international and domestic competitiveness. Thirdly, it examines the effects of international environmental measures on Sri Lankan export competitiveness.

Environment Management and Trade

It is widely held that the links between the environment and trade are at best tenuous. While it is production and consumption with the associated transport that causes environmental damage, the resulting management of the environment has implications for trade. Environmental degradation has been identified as arising from four major sources, namely, material source pollution, process pollution, product pollution and residual or waste pollution (Walter, 1975). Environmental management involves costs and these will be reflected in the prices of exportables and importables. How this affect the international flow of goods and services will depend on a host of factors. Environment management problems are highly industry and product-specific. Countries vary in assimilative capacity of their environments, as well as in the priorities to be accorded to environmental protection, the dependence on exportables and impotables, the type of products produced, and in their resource endowment, to name of the important factors.

While the environment is a global concern, the symptoms and causes of environmental degradation appear to be different as between the developed and developing countries. The problems of the developing countries have been identified as deforestation, desertification, demographic pressures, soil erosion and water pollution. In the developed world it is one of 'acid rain, atmospheric pollution, contaminated water, environmental mutagens and carcinogens, and toxic wastes'. Problem of global warming and depletion of ozone layer would appear to be more widespread or global in their impact and both the developed and developing countries as contributors towards it. The proximate causes would appear to be the poverty of the developing countries and the wealth of the developed countries. Given, however, the accelerated levels of economic activity, especially in the third world, which in turn is expected to be associated with increased environmental despoliation and an increasing demand for environmental management, countries could be expected to pay more attention to problems of environmental management and divert more resources towards it. This will affect international trade and this cannot be expected to be symmetrical.

Economists view environmental degradation as arising from the failure of the market to internalise the environment externalities. To the extent that the externalities are not reflected in relative costs, a difference in private and social cost and benefits arise. Paradoxically, public policy is required to redress the imperfections caused by market failure. Otherwise, the allocation of resources is such that the social optimum is not realised (Jogaratnam, 2000).

The potential impact of enforcing rigorous environmental standards on given country's exports and imports depends, however, on several factors. Firms may reallocate resources in country or relocate in resources in another country where environmental standards are less rigorous and where the assimilative capacity is greater. Imports may be restricted because of the inability of the exporting country to meet the required standards, or conforming to the standards raises cost in the exporting country. Much will depend on the extent to which the international community can harmonise standards. The fear is that the developed world would enforce much more rigorous standards resulting in looking the competitive edge. This may result in affected firm of 'Green Protectionism' while on the other hand, it is feared that lower standards and lax enforcement in the poorer countries could lead to 'ecological dumping'.

The available empirical evidence, however, suggests that the costs of environmental management do not significantly affect trade. A study by Krissoff *et al.*(1996) for the USDA cites numerous studies in support. Drawing on a voluminous literature, it addresses issues more specific to agriculture. A study in the early 1970s is reported to have found environmental costs for U.S. exports at 1.75 per cent of their total value and for U.S. imports at 1.5 per cent of their total value (Walter, 1975). A more recent study is reported to have estimated environmental regulation costs at about 1.85 per cent of total costs for 40 industries and between 1.85 and 2.89 per cent of total costs for 24 industries (Tobey, 1990). Additionally, it is also reported that reductions in output caused by environmental control costs are insignificant on average, although significant for some individual sectors, and that there is little evidence of any significant effect on patterns of trade. The envisaged developed country-developing country tensions over environmental regulations also appear exaggerated. The same cites data to illustrate that developing countries adopting developed countries themselves adopting stricter standards had little effect on production and trade in the nonferrous and energy sectors. It is also reported that Mexico's exports would be reduced by less than two per cent if it adopted an environmental tax equality which has improved in many countries, the effects on production trade and competitiveness remains small.

The trade effects of environmental policies on agriculture does not appear to be as adequately researched, as in the case of the other sectors, but it is felt that the above generalisations would apply (Krissoff *et al.*, 1996). However, it is also pointed out that proposed agro-environmental regulation in some countries, especially developed countries, could affect production costs, output and trade. A case in point is the potential impact of banning methyl bromide for soil and product fumigation, or banning of pesticide itself to the extent that alternative or more expensive, production losses could be associated with higher prices to consumers. Both export and imports would be affected in different countries depending on their quarantine regulations. Attempts to regulate nitrate levels in ground and surface water in the European Union, Japan and Taiwan, through controls on animal waste, could affect production and costs of livestock products (Krissoff *et al.*, 1996).

Impact on Environment Policy on Trade

There are two major views on the relationship between environmental regulation and international competitiveness. It is claimed that stringent environmental regulations impose significant costs on the domestic firms and industries reducing their international competitiveness in terms of declining exports, increasing imports compared with those from countries which have lower environmental standards and regulations. The loss of competitiveness is said to be greater in the so-called 'pollution-intensive' industries. Another view is entirely opposite to this, mainly due to Porter (1991) which suggested that environmental regulations lead to productivity improvements because a more cleaner environment leads to innovations due to stimulating effects of such regulations. Furthermore, countries adjusting early and investing in environmentally friendly technology would be able to create comparative advantage in these environmentally sensitive industries. Compliance with higher environmental regulations could also be an additional source of structural adjustments in particular for developing countries, as these regulations will promote less pollution-intensive industries (Sorsa, 1994).

There is a large body of literature, which uses neoclassical trade models to examine the impact of environmental regulations on competitiveness. (e.g. Copeland and Taylor, 1991). Although comparisons of these studies are difficult due to different methodologies used, Dean (1992) has summarised three major results of these studies. First, on average ECC is a smaller share of total costs of production. However, it has been shown that some individual industries such as construction, mining and plastics could suffer from loss of competitiveness due to high ECC. Second, generally, reductions in output caused by abatement costs are insignificant although some individual sectors have been affected. Third, the patterns of trade have not been affected significantly by the costs associated with environmental regulations.

Sorsa (1994) analyses trade flows in environmentally sensitive industries and ECC in seven industrial countries (Austria, Finland, Norway, Sweden, Germany, Japan and the United States). All these countries are clamed to have high environmental standards. First Sorsa finds that there is no systematic relationship between higher ECC and competitiveness in environmentally sensitive industries. For Germany, Japan and the U.S., the correlation coefficients between changes in exports share and ECC are not significant while that for Austria is positive. Second, in general countries under studies have been able to maintain competitiveness in environmentally sensitive products all through the period covered.

Local Environmental Regulations and Competitiveness

Some taxes, charges and standards designed to internalise environmental costs affect trade. It has been claimed by the business community that the goods and services subject to these levies become less competitive in the world markets compared with those sources from the countries which do not impose such taxes on business. However, as discussed before, there is no evidence to prove that this claim is true. As regards imports, generally these taxes do not become barriers to trade as long as they are applied to both imports and domestically produced goods at the same rate. In the case of Sri Lanka, the major environmental measure affecting competitiveness of local industries are:

1. Environmental Protection Licence (EPL) Scheme, and
2. Environmental Impact Assessment (EIA) Scheme.

Under EPL, an existing firm is required to obtain a licence to pollute from CEA which in turn issues a licence for a fixed fee, specifying the effluent standards that must be maintained by the firm. Currently, by and large these effluent standards are on water pollution. The EIA programme is applied to all new large industrial projects, under which firms are required to go through an approval process for environmental impact assessment by the CEA. In this analysis, two methods are used to examine whether the compliance cost associated with various environment standards and regulations has any effects on trade.

Compliance Costs

The estimation of compliance costs due to existing effluent charges is not an easy task for various reasons. A countrywide survey of industries is needed to assess the costs associated with such charges. A study by the UNDP has found that the capital costs of complying with the existing effluent charges are more than US $69 million in Sri Lanka. This number should be treated with caution as it was estimated subjected to a number of speculative assumptions. The breakdown of this estimate by industry is given in Table 9.1.

All in all, in the case of Sri Lanka, judging the available evidence from textile, leather, paper and rubber industries, one can suggest that the environmental compliance costs are not significant in total costs of production of most industries. This is mainly due to the fact that many firms do not abide by the required environmental standards and regulations.

Table 9.1

Capital Cost of Effluent Standards for Wastewater Treatment

Industry/Product	*Capital Cost (US $ Million)*
Textile	14.6
Food and Beverages	8.9
Tanning	4.7
Desiccated Coconut	4.5
Metal Processing	7.4
Rubber	16.6
Paints and Chemicals	2.0
Miscellaneous Products	9.9

Source: Industrial Policy Review of Sri Lanka, 1995.

The Analysis of International Competitiveness

The question being examined in this section is 'do domestic environmental standards and regulations lead to loss of international competitiveness of industries?' A well known concept, revealed comparative advantage (RCA), is used as a measure of international competitiveness to examine whether such regulations have affected adversely the competitiveness of Sri Lankan industries. In the absence of appropriate index showing the degree of stringency of environment regulations, industries are classified into Environment Sensitive (ES) or damaging and Environmentally Non-sensitive (ENS) categories to see whether ES industries have experienced any loss of competitiveness over time. The ES industries are assumed to have relatively high compliance costs resulting from environmental regulation and standards.

Many studies have used the concept of RCA to measure a country's relative export performance in individual product category. The index takes the values between zero and infinity. A RCA value greater than unity indicates that the country has a comparative advantage in that commodity. Higher the value of RCA, the greater is the country's comparative advantage in the commodity concerned.

The analysis is conducted on major Sri Lankan exports using data from International Trade Statistics of the United Nations. Major exporting industries are classified into ES and ENS using Central

Environment Authority of Sri Lanka classification where thirty-one industries and services have been listed as high polluting. It is clear from the Table 9.2 given below that, in general, the ES products have not experienced any loss of international competitiveness compared with ENS products.

International Environmental Regulations and Competitiveness

Another type of environmental regulations which have been designed recently are eco-labelling, packaging, eco-standards and recycling requirements. The purpose of these is to internalise externalities and protect the environment. Eco-labelling can also take form of either 'single issue labels' or 'lifecycle' type. The first type is used for managing environmental quality such as energy efficiency in refrigerators while the second type of the labels are designed for 'cradle to grave' management of products. Similarly, eco-packing schemes are being used to encourage recycling, reuse and proper disposal of the waste. In addition, some countries use eco-standards for environmental management. For example, products such as pesticides, metals and foods are subjected to a variety of standards in many countries.

Although most standards based on products can be justified in terms of the WTO Agreement on TBT and they are much better than the use of trade measures for environmental purpose, these measures have a great potential for misuse, acting as new trade barriers to trade. In particular, these measures can be used to favour domestic producers over importers and are likely to violate GATT/WTO principles. For example, it has been claimed that the German Packaging law introduced recently imposes enormous costs on exporters to the German market in relation to local German firms. Therefore, it is important for countries involved in trade to work towards mutual recognition and harmonisation of such measures. These measures are being used by many countries including Europe, North America and certain Asian countries, which are important destinations of Sri Lankan exports. Analysed below are the environmental-related barriers facing Sri Lankan exporters, based on some case studies, including the various environmental barriers facing Sri Lankan exports in other countries, in particular, Europe.

Table 9.2

Estimates of Comparative Advantage by ES and ENS Products Categories

Product	*ES/ENS*	*1987*	*1988*	*1989*	*1990*	*1991*	*1992*	*1993*	*1994*
Tea	ENS	382.5	414.4	343.1	335.1	305.1	220.5	214.8	239.2
Fruits	ENS	6.1	3.7	5.6	4.6	4.3	4.6	2.8	2.7
Spices	ENS	34.2	55.6	60.1	50.2	51.4	41.8	41.0	33.8
Tobacco	ES	0.6	1.0	1.9	1.4	1.8	10.0	9.2	8.8
Natural Rubber	ES	41.4	36.5	32.2	33.7	27.4	24.1	20.5	17.8
Charcoal	ES	107.8	59.3	44.2	22.9	22.1	12.4	3.3	9.2
Textile	ES	6.2	6.4	6.8	2.9	3.9	5.0	4.0	5.3
Mens' Outwear	ES	14.8	14.8	12.6	12.9	12.3	12.1	11.6	11.5
Women's Outwear	ES	2.3	19.6	25.9	15.9	16.6	20.2	19.2.	19.2
Undergarments	ES	20.6	18.1	24.2	18.9	17.4	20.4	19.8	17.4
Vegetable Oils	ENS				3.7	0.4	0.7	0.6	0.7
Chemical Products	ES	0.8	1.1	0.9	1.0	1.0	0.9	0.8	0.7
Rubber Articles	ES	0.01	0.3	0.2	2.4	2.5	2.8	3.1	2.8
Wood Manufacture	ES	1.5	2.6	0.6	2.1	4.1	2.2	2.4	2.5

Textiles

There are a variety of environmentally-related legislations covering textile. Table 9.3 given below presents a list of them applicable to textile imports to some European countries including Germany, the Netherlands and Sweden. One of the good examples of such regulation is Azo-dyes used in textile and leather production. This is proved to be harmful to human skin and banned in some Western countries (Germany and The Netherlands) and some countries are planning to ban such dyes and textiles made using such dyes.

Eco-labels

Voluntary eco-labels can be used by any textile-producing firms to distinguish its product as 'green' from other similar products. There are many eco-labels in the field including Oko-Tex 100, Nordic Swan and the Eco-label (EU-flower). Some of these are 'health labels' and some others are for cleaner production. There are also some organic labels (IFOAM, KRAV, and EKO) to prove that they are produced organically.

Food Products

Europe is the biggest market for exports of food products such as processed fruits and vegetables, edible nuts, herbs and spices, coffee and tea from developing countries. Trade related environmental measures are becoming increasingly important in these markets. Some of these environmental legislation applicable to imports in European countries such as Germany, The Netherlands, and Sweden are given in Table 9.4.

In addition to legislations and eco-labels, there are various other types of environmental requirements, which are useful for exporters from developing countries. These include the use of environment management system, such as ISO 14001, the use of waste management policy (e.g., packing requirements) and cleaner production (e.g., the use effluent treatment facility). Although countries can set restrictions only on imported products (according to WTO provisions), there is a growing concern about the production process (e.g., the use of chemicals) and it is becoming an important marketing factor in many Western countries. In the years to come, exporters from developing countries will find it difficult to penetrate to these markets if their

Table 9.3

Overview of Direct Legislation, Relevant to Textile Products

	EU	*Germany*	*The Netherlands*	*Sweden*
Certain Azo-dyes	Possible future legislation	Prohibition	Prohibition	Possible future legislation
Formaldehyde	Possible future legislation	Compulsory labelling above 1500 ppm.	Possible future legislation	Possible future legislation
Pentachlorophenol (PCP)	Possible future legislation	5ppm	5ppm	Possible future legislation
Nickel	Skin contact: 0.5/ug/cm2/week	Skin contact: 0.5/ug/cm2/week, labelling	Future legislation	Skin contact: 0.5/ug/cm2/week
Certain Flame Retardant	Prohibition	Prohibition	Prohibition	Prohibition
Cadmium	Paint/stabiliser: 100ppm	stabiliser: 100ppm	Paint/stabiliser: 100ppm	Colouring agents or stabiliser: prohibition
Asbestos	Prohibition	Prohibition	Prohibition	Prohibition
Plycholorobiphenyles and Terphenyles	Prohibition	Prohibition	Prohibition	Prohibition

Source: Environmental Quick Scan, CBI, The Netherlands

Table 9.4

Overview of Direct Legislation, Relevant to Food Products

EU Legislation	
Pesticides Residues-maximum Residue Levels, Chemical Contaminants	On-going programme of setting harmonised standards.
Food Additives	Legislation under development.
Microbiological Standards	Introduction of microbiological standards additional to standards for animal products under consideration.
Radiological Contamination	Maximum permitted levels.
Irradiation	Directive under development/Prohibition
Food Packing	Specific Legislation.

Source: CBI, The Netherlands.

goods are not produced using environmentally friendly production methods.

Concluding Remarks

It is clear now that trade-related environmental measures are becoming increasingly important in most developed countries. It is difficult to say how much Sri Lankan exporters are affected by the environmental instruments since there is no easy way to calculate the costs associated with these measures. However, the early adjustments to such measures will give a 'first mover' advantage to Sri Lankan exporters in penetrating these environmentally-sensitive markets. Not only can Sri Lankan continue to export the goods in which the country has comparative advantage but also create comparative advantage in new products related to environmentally-sensitive sectors.

As a country which has embarked upon an export-led industrial strategy as its development path and shown increasing concern over environmental protection, Sri Lanka can benefit from adjusting early and investing in environmentally-friendly technology which would create comparative advantage in the environmentally-friendly exports. Judging from the available evidence one can suggest that the domestic environmental compliance costs are not significant in total costs of production of most industries in Sri Lanka.

This is mainly due to the fact that many firms do not abide by the required environmental standards and regulations. The analysis conducted using the RCA concept, as a measure of international competitiveness, shows that international competitiveness of Sri Lankan industries has not been adversely affected by domestic environmental regulations. Various international environmental factors facing Sri Lankan exports in other countries, in particular, Europe have been examined and it is found that trade-related environmental measures are becoming increasingly important in most developed countries.

The early adjustment to these measures will give a 'first mover' advantage to Sri Lankan exporters in penetrating these environmentally sensitive markets. It is also important for countries involved in trade to work towards mutual recognition and harmonisation of such measures. Overall, it has been shown that the gains from trade more than compensate for environment damage arising from such trade. Beneficial effects of trade are enhanced when optimal environmental protection measures are in place.

References

Arrow, K.B. (1995). "Economic Growth, Carrying Capacity and Environment", *Science* (April).

Bardhan, P. (1995). "Research on Poverty and Development Twenty Years After Redistribution with Growth." in, M. Bruno and B. Pleskovic (eds.), *Annual World Bank Conference on Development Economics*. The World Bank, Washington, D.C.

Chenery, H. (1974). *Redistribution with Growth*, Oxford University Press, New York.

Chichilnisky, G. (1994). "North-South Trade and the Global Environment", *The American Economic Review*, 84(4), pp. 851-879.

Cooper, R.N. (1995). "Is Growth in Developing Countries Beneficial to Industrial Countries?" in, M. Bruno and B. Pleskovic (eds.), *Annual World Bank Conference on Development Economics*. The World Bank, Washington, D.C.

Copeland, B. and M. Taylor (1991). "North-South Trade and the Environment", *Quarterly Journal of Economics*, p. 109.

Dean, J. (1992). "Trade and Environment: A Survey of Literature", in Low (ed.) *International Trade and Environment, World Bank Discussion Paper 159*, World Bank, Washington, D.C.

Fishlow, A. (1995). "Inequality, Poverty and Growth: Where Do We Stand?" in, M. Bruno and B. Pleskovic (eds.), *Annual World Bank Conference on Development Economics*, The World Bank, Washington, D.C.

Jogaratnam, T. (2000). "Trade and Environment", *Sri Lanka Journal of Agricultural Economics*, 3(1), SLEA, Sri Lanka.

Islam, N. (1997). "Income-Environment Relationship: How Different is Asia?" *Asian Development Review*, 15(1), pp. 18-51.

Krissoff, B. *et al.* (1996). *Exploring Linkages Among Agriculture, Trade and Environment: Issues for the Next Century*, ERS, USDA, Washington, D.C.

Low, P. (1992). "International Trade and Environment", *World Bank Discussion Paper 159*, World Bank, Washington, D.C.

MaGuire, M. (1982). "Regulation, Factor Rewards, and International Trade", *Journal of Public Economics*, 17, pp. 335-354.

Nordstrom, H. and S. Vaughan (1999). "Trade and Environment", *Special Studies 4*, The World Trade Organization, Geneva.

Porter, Michaele (1991). "America's Green Strategy", *Scientific American*, April, p.168.

Rauscher, M. (1997). *International Trade, Factor Movements and the Environment*, Oxford University Press, New York.

Siamwalla, A. (1991). "The Relationship between Trade and Environment with Special Reference to Agriculture", paper presented at the *Conference on Agricultural Sustainability, Growth and Poverty*, sponsored by the German Foundation for International Development.

Sorsa, P. (1994). "Competitiveness and Environmental Standards: Some Exploratory Results", *Policy Research Working Papers*, World Bank, Washington D.C.

Tobey, H. (1990). *The Impact of Domestic Environmental Policies on Pattern of World Trade: An Empirical Test*, Kyklos, 43, pp. 191-209.

United Nation Environmental Programme (1999). *Trade Liberalisation and the Environment: Lessons Learned from Bangladesh, Chile, India, Philippines, Romania and Uganda*, United Nations, New York and Geneva.

Walter, I. (1975). *International Economics of Pollution*, Macmillan Press, London.

Part III

Policy Lessons

10

Environmental Trade Barriers and South Asian Exports: Lessons for National, International and Regional Action

NAGESH KUMAR
SACHIN CHATURVEDI

The Context and Introduction

The case studies of different South Asian countries prepared as a part of this project and other literature and experiences that have become available bring out a number of issues. It is clear that the emerging ETBs as they have been broadly defined have become a very potent trade barriers threatening a substantial proportion of South Asian exports. The WTO Agreements on SPS (Sanitary and Phytosanitary Measures) and TBT (Technical Barriers to Trade) aim to ensure that these standards and regulations are not used for protectionist purposes and do not cause adverse impacts on trade. However, at present there is considerable discretion available to importing countries to impose their own rules regarding these standards and other regulations such as inspection of imported products, specific treatment or processing of products, fixing of minimum allowable levels of pesticide residue, labelling and packaging requirements, good manufacturing practices, etc. The flexibility in these agreements has been exploited by developed countries to impose stringent environmental norms and standards that are acting as significant barrier for exports of developing countries. For instance, the SPS Agreement allows importing countries to impose SPS norms that are higher than international standards, provided there is a scientific basis. This provision has been used by various countries to check market access of specific countries. A case in point is unmanufactured tobacco where Japan insists on a DDT residue level of 0.4 particles per million (PPM) while the international standard is 6 PPM. Indian tobacco has a DDT residue level of 1-2 PPM which is well below the internationally permissible levels. But Japan does not allow tobacco imports from India on phyto-sanitary grounds. Very often flimsy grounds and minute risks assessments are used to justify imposition of these higher standards. For instance, adoption of a new aflatoxin standard in the European Union that would reduce health risk by approximately 1.4 deaths per billion a year, as reported in chapter 1. While there are not a billion people in the whole of EU, African exports of cereals, dried fruits and nuts to EU are expected to decline by 64 per cent as a result of adoption of these standard with an estimated loss of export earnings of the order of US$ 670 million (Jha and Vossenaar, 2001). Even though international standards such as Codex are evolved by developed countries, they themselves do not adopt them.

Another problem is the wide variation in the standards adopted by different importing countries. Different countries follow different norms of aflatoxins and pesticide residue increasing the compliance cost for the exporting countries. The European Union countries have gone as far as to adopt a single currency. However, their standards with respect to pesticide residue vary widely; for instance, from 0.01 PPM in the case of the UK to 0.03 in the Netherlands to 0.10 PPM in Germany for Aldrin and Dieldrin (Kithu, 2001). Many of these environment related standards are imposed in a less than transparent manner and sometimes are accompanied by other requirements such as Good Manufacturing Practice. There are cases of detention of consignments of Indian spices in Spain, Italy and Germany without any satisfactory explanation forthcoming on the changes made in their regulations on microbial contamination and pesticide residue despite transparency obligation under the Article 7 of SPS Agreement (Jha and Vossenaar, 2001).

Sometimes additional standards are just imposed informally by the importing countries on their exporters as pre-conditions for doing business, as the case study of grapes exports by India has demonstrated. Often importers use these standards to depress prices of imports. For instance, foreign buyers reportedly paid lower prices for leather products from Bangladesh because of the poor environmental conditions in the tanning sector in Dhaka (UNCTAD, 1998:145). The experience of India in the case of export of flowers to Japan is also similar (chapter 5).

Another concern that is clear is that the cost of compliance is substantial and often beyond the competence of many of the enterprises especially the smaller ones. A rough estimate for setting up a moderate lab for testing and analysing samples of spices comes to about four million rupees (Kithu, 2001). Furthermore, in a number of cases the technology for compliance with the standards may not be available easily in developing countries.

Many developing countries have experienced losses in exports because of difficulties in complying with these emerging standards and regulations. South Asian countries, in particular, are affected badly as some of their major items of their export interest such as textiles, marine products, spices, among others, are those affected by these barriers. There is an urgent need to respond to this challenge. Besides the policy responses at the national level, actions at the international and at the regional levels would be often necessary for mitigating the adverse effect of these emerging trade barriers on South Asian countries. In this chapter we provide an inventory of such policy responses to the national, international and regional levels.

Issues for Action at the National Level

It is quite clear that governments do have to play an important role because especially when the exporters are small and medium enterprises and it may be beyond their capacity to comply with the new standards.

Information Sharing

Governments should create mechanisms for disseminating information on the new emerging standards adopted by different countries to the exporters especially the SMEs. This information dissemination could help in reducing the rate of rejections of export consignments and thus help avoid the costly process of bringing them back or destroy them.

Capacity Building for Compliance

Given the high cost of compliance which may be beyond the capacity of smaller enterprises, it may be of critical importance to create common infrastructural facilities such as test laboratories with international accreditation. The government agencies may assist the smaller enterprises in compliance with the new standards through provision of technology and necessary raw materials. The capacity building could also include organisation of extension programmes which cover training of entrepreneurs, managers and different levels of workers in compliance of the required standards. Governments could also facilitate formation of cooperatives and self-help groups of exporters so that they could share costs and create common facilities such as test laboratories, water treatment plants, affluent treatment plants besides information dissemination. The patterns of clustering or agglomeration of exporters specialised in one product in certain areas that exist in the region (e.g. knitwear makers in Ludhiana) suggest that formation of these self-help groups/ cooperatives could be effective. Furthermore, the governments have a role to play in ensuring the availability of all necessary raw materials for compliance with specific standards. For instance, the Nepal case study shows that the non-azo-dyes for the carpet industry are not available in the country thus impeding the compliance with the ban on azo dyes imposed in the European Union.[1] So governments have to make sure that substitutes of azo-dyes are available to the carpet manufacturers. The relevant governments could consider setting up Exporters Assistance Centres as one-stop shops to cover all different tasks such as dissemination of information, creation of test facilities, the provision of technical assistance and training, etc.

Evolve National Standards

As a result of the liberalisation of trade regimes, the South Asian economies have become increasingly open to import of goods from abroad. In particular, a lot of imported fruits and vegetables are finding their way to the South Asian markets. There is a fear that

1. See UNCTAD 1998, for more details.

these imports might become carriers to new pests and other harmful germs to the region. It is important for the governments to evolve their own environmental standards and quarantine regulations which should be strictly enforced.

Effective Participation in International Standards Setting

The experience of past years suggests that the international standards such as those in Codex Commission are evolved with very little participation of developing countries. Therefore, what developed countries propose gets approved as a default. Developing countries need to participate in these events effectively with their preparation of implications and other issues and concerns of alternative standards.

Generating Consciousness about Environmental Concerns in the Industry

The governments can play an important role in generating the consciousness of environmental considerations in industry through media and other means. The Indonesian example of colour grading of the companies on the basis of the environmental responsibility is a case in point (chap. 7). The increased environmental consciousness generated through the media of this scheme has led to significant decline in market valuation of the companies that were labelled as relatively poor (black) in terms of environmental standards. It suggests that such schemes could build considerable pressure on companies to become environmentally more conscious.

Databases of Experiences

The field-work conducted as a part of the country studies brings out a variety of experiences of imposition of environmental standards by importers on exporters of the countries studied, sometimes in an informal manner and often under the cover of good manufacturing practices. It is important to compile a database of such experiences so that as to verify if certain regulations imposed are not consistent with SPS and TBT norms and hence necessary redressal could be sought through WTO Dispute Settlement.

Exploiting Opportunities for Environmentally-friendly Goods and Services

The South Asian countries should take advantage of the growing environmental consciousness in the developed world by pushing exports of the environmentally friendly goods and services. The South Asian countries have a natural advantage in organic farming which should be exploited fully. The region has used vegetable dyes such as indigo for centuries in the textile industry. It may be fruitful to create a brand in the Western world on the basis of use of vegetable dyes. Similarly, exports of herbal medicines, cosmetics, eco-tourism among other environmentally friendly goods and services could be pushed. Bangladesh has rapidly got an eco-label for its jute-packaging bag which has enhanced the sales of these bags significantly (UNCTAD, 1998).

Diversification of Export Commodities/Markets

It is clear from the country studies that South Asian countries are heavily dependent on exports of few commodities and on few markets; for instance, Nepal's dependence on carpet exports, Bangladesh's dependence on garments, marine products and leather goods. Even though India's exports structure is more diversified, the share of textiles and garments in India's exports is quite substantial at 33 per cent. Similarly, there is a high dependence on specific markets. For instance, Germany accounts for 80 to 90 per cent of Nepal's exports of carpets. The dependence on few products and markets for exports make these economies highly vulnerable. Therefore, in medium to long term, the objective of the government policy should be to diversify the composition of exports as well as the coverage of markets.

Equivalence Agreement

As part of the capacity building exercise it is also important to establish equivalence agreements with major trade partners. These agreements are also being encouraged in the SPS and Codex framework. The equivalence agreements may also be extended at the regional level for further expediting the technical cooperation and overcoming the constraints of export rejections (chap. 6). They would also help in converging the various approaches being used for technical assessments (chap. 6).

Issues for International Action

These include a number of issues which have to be taken up at the WTO and other international fora.

Effective Technical Assistance

Both SPS and TBT Agreements in WTO contain provisions for technical assistance to be provided by the importing countries to assist the exporters in their compliance with the new standards imposed by them. The SPS agreement under Article 9 includes:

> "Provision of technical assistance to other Members specially developing country Members ... in the areas of processing technology, research and infrastructure, including in the establishment of national regulatory bodies and may take the form of advice, credits, donations and grants including for the purpose of seeking technical expertise, training and equipment to allow such countries to adjust to and comply with sanitary and phyto-sanitary measures necessary to achieve the appropriate level of sanitary and phyto-sanitary protection in their export markets."

Furthermore, the SPS Agreement provides:

> "Where substantial investments are required in order for an exporting developing country Member to fulfill the sanitary and phyto-sanitary requirements of an importing Member, the latter shall consider providing such technical assistance as will permit the developing country Members to maintain and expand their market access opportunities for the product involved" (Art. 9.2).

Similarly, the TBT Agreement under Article 11 provides for advice and technical assistance, if requested, on mutually agreed terms and conditions. However, there is little evidence of the promised technical assistance forthcoming. Following the ban imposed by European Commission on exports of fisheries products originating in Bangladesh in 1997 in the light of deficiencies in the infrastructure and hygiene fisheries establishments in the country, Bangladesh Frozen Foods Exports Association had sought technical assistance from European Union experts. That assistance was provided but the association had to pay for it from its own resources (UNCTAD, 1998). It is clear, therefore, that the obligation for technical assistance has not been fulfilled by developed countries in an adequate manner.

Reviews of SPS and TBT Agreements

Developing countries may consider seeking reviews of the Agreements on SPS and TBT at the WTO Committee on Sanitary and Phyto-sanitary Measures and the Committee on Technical Barriers to Trade to reduce the discretion and flexibility available under these Agreements that is being misused, universal application of international standards, and to bring about greater transparency in the imposition of standards, good manufacturing practices and all other regulations.

Special and Differential Treatment (S&DT)

Like other WTO Agreements, S&DT in the case of SPS and TBT agreements has also been reduced to a longer transition period for developing especially the least developed country members. However, the longer transition period provided under these Agreements is of hardly any use because the importing enterprises start insisting on the new product standards due to high consumer consciousness for these standards. Therefore, the S&DT has to be of a more substantive type. It could be, for instance, in the form of additional financing (along with technical assistance) to cover the costs of compliance by the developing country exports of the new standards.

Compulsory Licensing of Technology

Sometimes the compliance of new standards is constrained by non-availability of technology due to strengthening regime of IPR protection. The TRIPs Agreement should provide for compulsory licensing on environmental grounds.

Efforts may be made to resolve the constraints emanating from different languages being used in issuing guidelines. For example, Spain insists on export related forms to be submitted in Spanish and moreover, the instructions issued are also in the Spanish language.

Assistance for Effective Participation of Developing Countries in Setting of International Standards

As observed earlier, the international standards are evolved without much participation of developing countries. The international community should subsidise the participation of developing countries in the meetings of international standards setting bodies besides assisting them in their technical preparations for their effective participation in these fora.

Issues for Regional Cooperation

There are a number of areas where a regional approach could be fruitful as we see below.

Coordination in International Negotiations

The countries of South Asian region export many products that are common to all. Hence their concerns are quite similar. A region-wide coordination in international standards setting events and in the WTO Committees on SPS and TBT and other negotiations would be serving their interest better.

Cooperation in Compliance with the Emerging Environmental Standards

The regional cooperation could be effective in sharing costs of compliance with the emerging environmental standards. The regional cooperation could cover joint development of products which meet the new regulations and hence sharing the costs. In this context, a case in point is a highly successful project for development of aflatoxin risk free groundnut jointly conducted by the Indian Council of Agricultural Research and the UNDP. This project successfully brought down the aflatoxin levels to 0-5 PPB in 80 per cent of the samples at the end of three years term of the project in a high risk area of Andhra Pradesh, India against the permissible limit of 15 to 20 PPB in developed countries such as Australia, Canada, USA (Basu and Radhakrishnan, 2001). Such projects could be fruitfully undertaken at the regional level. The regional cooperation could also cover creation of regional institutional infrastructure such as test laboratories where the costs are high. The geographical contiguity in the region would facilitate the optimal utilisation of such infrastructure. Joint training programmes and other measures to build local capacity would also be fruitful.

Regional Eco-label Programmes

A number of South Asian countries including India have tried to launch their own eco-label programmes. However, these programmes have had limited success. Another area of regional cooperation could be promotion of regional eco-labels which might be more acceptable and popular than national eco-labels.

Evolving Regional Standards for Products of Regional Export Interest

Another area of fruitful regional cooperation could be to evolve regional standards especially for products of our export interest and have them accepted globally. Similarly, it would be fruitful to define environmentally sensitive goods from a regional perspective.

Concluding Remarks

In this chapter we have put together an inventory of policy lessons emerging from the analysis of the previous chapters. These policy lessons cover some issues for action at the level of national governments that are likely to prepare the exporters in these countries face the emerging challenge more effectively. Then there are issues to be addressed at the international or WTO level. These will help moderate the incidence of these new standards on the South Asian countries. Finally, we have identified some areas in which regional cooperation would be fruitful.

References

Basu, M.S. and Radhakrishnan (2001). "Capacity Building of Dryland Groundnut Farmers in Addressing SPS Issues–A Success Story", presented at the UNCTAD Workshop on *A New WTO Round Agriculture, SPS, and the Environment Capturing the Benefits for South Asia*, New Delhi, January.

Bhatia, Ujal Singh (2001). "Promotion of Organic Agricultural Exports in India: Opportunities, Constraints & Policy Issues", presented at the UNCTAD Workshop on *A New WTO Round Agriculture, SPS, and the Environment Capturing the Benefits for South Asia*, New Delhi. January.

Dasgupta, Amit (2001). "Agreement on Sanitary and Phytosanitary Standards: SAARC Concerns", presented at the Workshop *A New WTO Round – Agriculture, SPS, and the Environment Capturing the Benefits for South Asia*, SAARC Secretariat, Kathmandu, Nepal.

Henriksen, Anna, Mai Sugiyama (1998). *Voluntary Environmental Labelling and the World Trade Organisation*, International Trade, Environment and Development, Institute of Economics Copenhagen University.

Jha, Veena and Rene Vossenaar (2001). "Strengthening Developing Countries' Capacities to Respond to Health, Sanitary and Environmental Requirements", presented at the UNCTAD Workshop on *A New WTO Round Agriculture, SPS, and the Environment Capturing the Benefits for South Asia*, New Delhi, January.

Jha, Veena (2001). *Strengthening Developing Countries Capacities to Respond to Health, Sanitary and Environmental Requirements: A Scoping Paper for South Asia*, UNCTAD, New Delhi.

Kithu, Charles. J. (2001). "Issues on SPS and Environmental Standards for India", presented at the UNCTAD Workshop on *A New WTO Round Agriculture, SPS, and the Environment Capturing the Benefits for South Asia*, New Delhi, January.

TCES (2000). *Environmental and Other Labelling of Coffee: The Role of Mutual Recognition Supporting Cooperative Action*, Terra Choice Environmental Services Inc., Ontario.

UNCTAD (1998). "Trade and Environment and the LDCs", in *The Least Developed Countries 1998 Report*, United Nations, New York and Geneva, pp. 137-155.

————. (2002). *Actions Taken by UNCTAD in Assisting Developing Countries to Implement Agenda 21*, UNCTAD Secretariat.

Wilson, John S. (2001). *Global Trade and Food Safety: Winners and Losers in a Fragmented System*. Development Research Group, The World Bank, Washington, D.C.

Index

A

Agenda 21 30
agriculture-sector/product/import 73, 76
aerosol 68
Africa 42
aflatoxin 21, 23, 42, 108
aid agencies 35, 36
appellate body 44, 107
APEC 48
Australia 22, 32, 113, 129, 137, 156, 226

B

Bangladesh 24, 42, 51, 103, 122, 211, 212, 214, 215–218, 221, 223, 224–226, 228, 229, 230, 250, 254, 255
Basel Convention 39
beef hormone 32
biodiversity 37, 169, 174, 178
BOD 166, 172, 177, 226, 228

C

Canada 44, 113, 129, 134, 137, 155
COD 166, 172, 179
cadmium 128, 190, 197-199
Caribbean 23, 44
Cartagena Protocol on Biosafety 44, 126, 129
capacity building 50, 156, 252
certification 22, 35, 41, 42, 131, 133, 135, 138,•140
chemical industry 49, 50, 52, 123
China 110, 127, 137, 214
Clean Development Mechanism 127
Codex Alimentarius Committee/ standard 23, 155, 157, 161
Committee on Trade and Environment 20, 30, 62, 102
competitive advantage 41
comparative advantage 33, 235, 237, 253, 254
compliance cost 45, 51, 54, 108, 237
common effluent treatment 116, 225
Convention on Biological Diversity 129
Convention on Trade in Endangered Species of Flora and Fauna 30, 39, 65, 67, 69, 72, 76, 77
crocodile farming 39
cyanide 22
CFC 29, 68

D

DDT 113, 249, 250
dirty industries 37, 45, 46, 48, 49, 117, 123
Dispute Settlement Mechanism 43, 131
durian 32
dye/azodyes 51, 166, 167, 168–170, 172, 173

E

eco-mark 133, 135

eco-labelling 42, 43, 131, 132, 135, 138, 169, 170, 173, 187, 239
ecosystem 37
eco-standard 51
effluent standards/treatment plant 182, 183
environment impact assessment 237
environmental measures 45, 53, 101, 102
environmental management system 171
environment related non-tariff barriers 103
environmental trade barriers 19, 113
environmental requirements 24, 34, 108
environmentally friendly products 52, 53, 54, 62, 133, 136, 138, 243
environmentally sound technologies 52–54, 130
environmentally sensitive goods/ products/commodities 20, 37, 38, 53, 62, 236
environmental (process) standards 20, 34, 101, 122, 211
environment related trade barriers/ regulation 103, 104, 105, 106, 107, 249
Europe 29, 31, 32, 43, 105, 241
European Union 21, 22, 23, 24, 31, 32, 33, 42, 62, 63, 64, 69, 70, 71,72, 73, 75, 76, 77, 79, 104, 108, 110, 113, 114, 115, 116, 119, 127, 128, 129, 130, 136, 137, 156, 160, 214, 215, 219, 220, 221, 222, 228, 235, 250

F

France 47, 113, 115, 116
food additives 21, 111, 112
Food and Agriculture Organization 223
food safety regulation 19, 102, 104
foreign direct investments 46, 47, 48, 122–126
free trade/trade liberalisation 40, 50, 53
fruits and vegetables 70

G

GATT 30, 32, 33, 40, 209, 231, 232, 239
genetically modified organisms 31, 44, 126, 129, 130
General System of Preference 215, 230
global warming 53, 169
Germany 21, 29, 31, 115, 114–116, 133, 136, 137, 166, 225, 239, 241, 250, 254
Guatemala 36

H

Hazard Analysis Critical Control Points 42, 103
harmonisation standard 40, 101
health standards 70
health mark 70
HS Classification 107
hygiene requirement 21, 42

I

India 21, 22, 29, 50, 51, 52, 53, 103-109, 111, 113-115, 122, 123, 125, 126, 127, 130, 136, 138, 140
Ireland 48
International Plant Protection Convention 222

J

Japan 107, 109, 110, 137, 214, 219, 235, 236, 250
jute 213, 215, 254

K

Kyoto Protocol 126, 127

L

labelling 43-45, 111, 131, 132, 135, 239

leather industry/export 51, 115, 116
like products 44
living modified organism 129

M

Malaysia 43, 44
market access 40, 50, 62–64
Marrakesh Agreement 232
Mexico 47, 48, 188
MFN 70, 77, 78, 126
microbial contamination 21, 22, 109
Montreal Protocol 29, 39, 53, 65, 67, 69, 169
Morocco 47
Mozambique 23
multinational companies 63
Multilateral Environment Agreement 29, 39, 53, 62, 65, 69
mutual recognition agreement 42
Myanmar 113

N

NAFTA 39
natural resources 62
net export 45, 49
Newly Industrialised Countries 209
non-governmental organisation 31, 33, 163
non-tariff measures/barriers 22, 24, 61-65, 69, 70-79, 106, 108, 110, 111

O

OECD 30, 31, 38, 46, 47, 48, 52, 111, 130, 132, 138, 187
organic food/farming 34, 52
ozone layer 34, 67, 233

P

packaging requirements 50, 51, 108
Pacific countries 23
Pakistan 22, 24, 53, 165, 166, 169, 170, 172, 177, 181-183, 185–187
pesticide residue (MRL) 21, 22, 50, 103
Philippines 40, 188
pollution haven 45, 46, 48, 49, 50
pollution intensive production/ industries 45-49, 54, 235, 236
pollution or pollution abatement cost 45, 49, 117
polluter pays principle 128
precautionary measure/principle 32, 33
product labelling 32
product standards 29, 44, 54
protectionism 61, 234
public goods 34
product and process methods 34, 35, 61

Q

quantitative restriction/regulations 108, 217

R

relocation of industries 45, 117
regional trading agreement 63
refrigerator 68
risk assessment 32, 154, 155
Romania 48

S

SAARC 49
salmon 43, 44
shellfish 51, 116
shrimp-turtle case 44
Southeast Asia 32, 38
South America 119, 120, 121
South Asia 21, 24, 25, 37, 38, 44, 63, 64, 79, 103, 111, 121, 122, 126, 130, 251, 252, 254, 258
SPS 19–23, 30, 35, 41-44, 101, 102, 107–109, 126, 129-131, 139, 153, 154, 157, 160, 222, 223, 249, 250, 255, 256
Spain 48, 256

special and differential treatment 256
SMEs 22, 62, 114, 115, 127, 138, 171, 251
Sri Lanka 21, 112, 232, 237–239, 243, 244

T

TBT 19, 30, 35, 41, 43, 44, 107, 129, 138, 239, 253, 255, 256
technology transfer 52
tea/industry 51
textile/clothing sector 33, 63, 114, 189, 218
Thailand 22, 32, 43, 44
trade agreements 48
trade restrictive 106
TRIPS 35, 256
turtle excluder device 44

U

Uganda 23
UK 115, 136, 137, 250
UNCED 30, 61, 62
UNCTAD 30, 62, 64, 71, 107
UNDP 108, 219, 229, 237
UNIDO 225
United Republic of Tanzania 23
Uruguay Round 22, 209
USA 32, 41, 42, 44, 45, 47, 64, 103, 107, 110, 111, 117, 119, 129, 130, 131, 155, 211, 214, 230, 234

V

Venezuela 47, 107

W

Word Bank 47, 218, 229
WTO 19–21, 35, 71, 101, 102, 107, 113, 126, 132, 139, 153, 154, 161, 209, 239, 249, 253, 255, 256, 258

Z

Zimbabwe 39